Carpenters
and
Builders
Library
No.1

by John E. Ball

THEODORE AUDEL & CO.
a division of

HOWARD W. SAMS & CO., INC.
4300 West 62nd Street
Indianapolis, Indiana 46268

FOURTH EDITION

FOURTH PRINTING—1978

International Standard Book Number: 0-672-23240-5
Library of Congress Catalog Card Number: 76-24079

Foreword

Carpentry, as a branch of the building trades, is one of man's oldest and most significant vocations. It encompasses all processes and degrees of construction with wood—from the simple sanding of a board for a cabinet shelf to the complex task of building an entire house. This book, the first in a four-book series, provides the knowledge necessary to attain success in any carpentry project attempted by the home owner, hobbyist, apprentice, journeyman, and master carpenter.

The text gives a complete explanation of the numerous and varied hand tools of which a carpenter must have a thorough understanding. The steel square, which is the "computer" of the carpentry trade, and its applications are discussed in great detail with the aid of typical examples and illustrations that show the multiplicity of its uses. Joints and the art of joinery are also examined in depth, along with some of the many applications of the various joints—cabinetmaking, furniture construction, patternmaking, etc.

The other three texts in this series provide a complete and thoroughly rounded range of topics which are of primary concern to both the would-be and the professional woodworker. These topics consist of such carpentry fundamentals as builders mathematics, surveying, architectural drawing, house construction framing, foundations, use of power tools, painting, millwork, and the construction of doors, windows, and stairs.

The purpose of this volume, then, is to present the reader with the concepts and basic rules governing the use of woodworking tools and the construction of joints, which are necessary for skilled workmanship in the carpentry and building trades. This purpose is accomplished through the use of hundreds of illustrations and an explanatory text that provide the all-important details in such a manner that this material will be of unlimited assistance to both the layman and the master carpenter.

JOHN E. BALL

Contents

patterns — cores — core prints — core boxes — draft — finish — shrinkage — blueprints — patternmaking joints — summary — review questions

CHAPTER 24

Kitchen layout planning — kitchen wall cabinets — false cabinet walls — cabinet installation — unit kitchen cabinets — summary — review questions

CHAPTER 1

Woods

Wood is our most versatile, most readily obtainable building material, and a general knowledge of the physical characteristics of the various woods used in building operations is extremely desirable for the carpenter. A mastery of this subject is only attainable through long experience.

Wood may be classified:

1. Botanically. All trees which can be sawed into lumber or timbers belong to the division called Spermatophytes. This includes softwoods as well as hardwoods.
2. With respect to its density, as
   a. Soft.
   b. Hard.
3. With respect to its leaves, as
   a. Needle or scale leaved, botanically Gymnosperms, or conifers, commonly called softwoods. Most of them, but not all, are evergreens.
   b. Broad-leaved, botanically Angiosperms, commonly called hardwoods. Most are deciduous, shedding their leaves in the fall. Only one broad-leaved hardwood, the Chinese ginkgo, belongs to the subdivision Gymnosperms.
4. With respect to its shade or color, as
   a. White or very light.
   b. Yellow or yellowish.
   c. Red.
   d. Brown.
   e. Black, or nearly black.

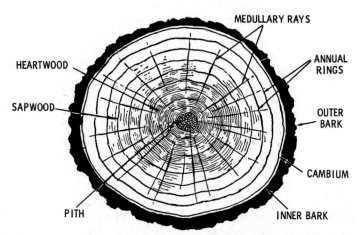

Fig. 1. Cross section of an oak nine years old, showing pith, concentric rings comprising the woody part, the cambium layer, and the bark. The arrangement of the wood in concentric rings is due to the fact that one layer was formed each year. These rings, or layers, are called annual rings. That portion of each ring formed in spring and early summer is softer, lighter colored, and weaker than that formed during the summer and is called spring wood. The denser, stronger wood formed later is called summer wood. The cells in the heartwood of some species are filled with various oils, tannins, and other substances, which make these timbers rot-resistant. There is practically no difference in the strength of heartwood and sapwood, if they weigh the same. In most species, only the sapwood can be readily impregnated with preservatives.

5. With respect to grain, as
   a. Straight.
   b. Cross.
   c. Fine.
   d. Coarse.
   e. Interlocking.
6. With respect to the nature of the surface when dressed, as
   a. Plain—example, white pine.
   b. Grained—example, oak.
   c. Figured or marked—example, bird's-eye maple.

A section of a timber tree, as shown in Figs. 1 and 2, consists of:

1. Outer bark—living and growing only at the cambium layer. In most trees, the outside continually sloughs away.

2. In some trees, notably hickories and basswood, there are long tough fibers, called bast fibers, in the inner bark. In some trees such as the beech, they are notably absent.
3. Cambium layer—Sometimes this is only one cell thick. Only these cells are living and growing.
4. Medullary rays or wood fibers which run radially.
5. Annual rings, or layers of wood.
6. Pith.

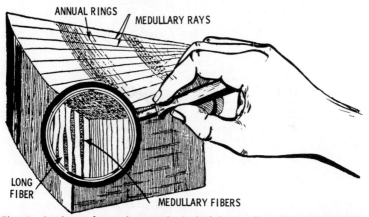

Fig. 2. A piece of wood magnified slightly to show its structure. The wood is made up of long, slender cells called fibers which usually lie parallel to the pith. The length of these cells is often 100 times their diameter. Transversely, bands of other cells, elongated but much shorter, serve to carry sap and nutriments across the trunk radially. Also, in the hardwoods, long vessels or tubes, often several feet long, carry liquids up the tree. There are no sap-carrying vessels in the softwoods, but spaces between the cells may be filled with resins.

Around the pith, the wood substance is arranged in approximately concentric rings. The part nearest the pith is usually darker than the parts nearest the bark and is called the heartwood. The cells in the heartwood are dead. Nearer the bark is the sapwood, where the cells are *living* but not *growing*.

As winter approaches, all growth ceases, and thus each annual ring is separate and in most cases distinct. The leaves of the deciduous trees, or those which shed their leaves, and the leaves of some of the conifers, such as cypress and larch, fall, and the sap in the tree may freeze hard. The tree is *dormant* but *not dead*. With the warm days of the next spring, growth starts again strongly, and the cycle is repeated. The width of the annual rings varies greatly, from 30 to 40 or more per inch in some slow-growing species, to as few as 3 or 4 per inch in some of the quick-growing softwoods. The woods with the narrowest rings, because of the large percentage of summer wood, are generally strongest, although this is not always the case.

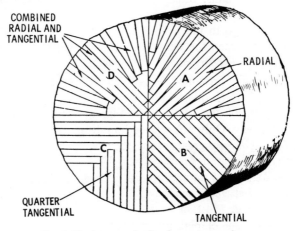

Fig. 3. Various methods of quartersawing.

LUMBERING

This consists of the operations which must be performed when preparing the wood for the carpenter. It includes these steps:

1. Logging.
2. Sawing.
3. Drying, dressing, and sizing.

Logging

This constitutes felling the trees, cutting the logs, and delivering them to the sawmill. In the large operations, the logs are taken out of the woods by means of aerial cableways, or roads are built and the logs are trucked out. Where waterways are convenient, the logs are often floated down to the mill. Practically always, they are dumped into a "log pond" for easy handling.

Drying, Dressing, and Sizing

The sawed lumber or timber may be piled on sticks and allowed to air-dry, at least partially, or it may be sent direct to the dry kilns, if time is too valuable. When the moisture content comes down to 12 to 19%, the lumber is said to be air-dry.

In the larger mills, often band-mills, the logs are sawed into bolts, with one dimension having the width of the lumber desired. The bolts then go to the gang saws, a battery of relatively small circular saws spaced the desired distance apart, and at one pass, the bolt is ripped to the rough size desired.

It is commonly presumed that in quartersawing a log, it is actually quartered. Each quarter is laid on the bark and ripped up as shown in one of the quarters in Fig. 3. Actually, it is rarely done in this way. Only a few wide boards are obtainable, and the waste is prohibitively great. More often, when a majority of quartersawed lumber is desired, the log is started, as illustrated in Figs. 4 and 5, sawed up until a good figure shows, then it is turned over and sawed up. Then there is no waste but slab, and the boards are wide. Most of the quartersawed lumber marketed is simply graded out of plain-sawed lumber. There is only a slight difference in the value of plain and rift-sawed lumber. In rift-sawed softwoods, grain which slopes no more than 45° is termed edge-

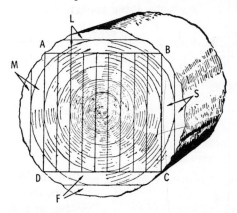

Fig. 4. Plain or bastard sawing, sometimes called flat or slash sawing. The log is first squared by removing boards MS and LF, giving the rectangular section ABCD. This is necessary to obtain a flat surface on the log.

grained, or quartersawed. Boards shrink most in a direction parallel with the annual rings. For this reason, door stiles and rails are often quartersawed.

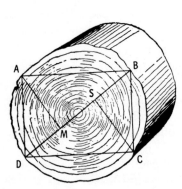 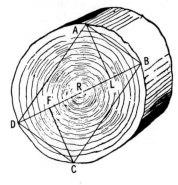

A. For one large beam, divide the longest diameter DB into three equal parts; erect perpendiculars at M and S, and join points thus obtained to form rectangle ABCD.

B. For one stiff beam, divide the longest diameter DB into four equal parts; erect perpendiculars at L and F, and join points thus obtained to form rectangle ABCD.

Fig. 5. Obtaining beams from a log.

DEFECTS

The defects found in manufactured lumber, as shown in Figs. 6 and 7, are grouped in several classes:

1. Those found in the natural log, as
   a. Shakes.
   b. Knots.
   c. Pitchpockets.

2. Those due to deterioration, as
   a. Rot.
   b. Dote.

3. Those due to imperfect manufacture, as
   a. Imperfect machining.

b. Wane.
c. Machine burn.
d. Checks and splits from imperfect drying.

Heart shakes, as shown in Fig. 6, are radial cracks that are wider at the pith of the tree than at the outer end. This defect is most commonly found in those trees which are old, rather than in young vigorous saplings; it occurs frequently in hemlock.

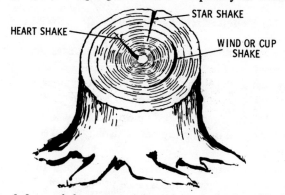

Fig. 6. Some of the various defects than can be found in lumber.

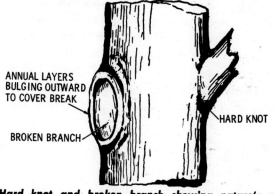

Fig. 7. Hard knot and broken branch showing nature's method of covering the break.

A wind or cup shake is a crack following the line of the porous part of the annual rings and is curved by a separation of the annual rings. A wind shake may extend for a considerable distance up the trunk. Other explanations for wind shakes are expansion of the sap wood and wrenchings received due to high winds (hence the name). Brown ash is especially susceptible to wind shakes.

A star shake resembles the wind shake but differs in that the crack extends across the center of the trunk without any appearance of decay at that point; it is larger at the outside of the tree.

Dry rot, to which timber is so subject, is due to fungi; the name is misleading, as it only occurs in the presence of moisture and the absence of free air circulation.

STACKING AND SEASONING OF LUMBER

In preparing lumber for the market, it is necessary that it be seasoned; that is, the moisture should be expelled. The more thoroughly this is done, the less likely is the lumber to shrink or decay. The methods may be classed as natural and artificial.

Natural seasoning consists of exposing the sawed lumber to the free circulation of air. The lumber is usually stacked in piles, as shown in Fig. 8.

Artificial drying, or kiln drying, is done when time must be considered and where good appearance is important. The strength of the lumber is not affected if the drying is done properly, and kiln-dried lumber is much brighter and freer from stain than air-dried lumber. There are many types of kilns in common use, and all of them are entirely satisfactory when correctly handled. Improper kiln drying, however, can result in serious checks, splits, and warpage, as shown in Fig. 9. For this reason, lumber should be dried before dressing and sizing.

There are two kinds of kilns that are normally used in air drying:

1. Natural draught.
2. Induced draught.

Because of the slow circulation of the air in natural-draught kilns, the air will become charged with moisture to a much greater degree than if it were forced through rapidly, as is done in the induced-draught kiln. In induced-draught kilns, the air, after passing over the lumber and becoming charged with moisture, circulates over cold condensing plates where it is relieved of its

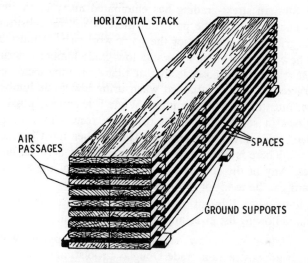

Fig. 8. Horizontal stack of lumber for air drying.

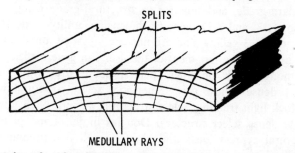

Fig. 9. A board with splits along the medullary rays. This condition is caused by too-rapid kiln drying.

moisture. After being reheated, it again passes over the lumber, each time abstracting more moisture from the lumber. Since the operating cost of the blower and condensing equipment is considerable, lumber to be dried in an induced-draught kiln should first be partially air dried.

SELECTION OF LUMBER

The selection of the grade of lumber to be used for any specific purpose is often, though not always, left to the discretion of the carpenter. Modern stress-grading has eliminated many of the uncertainties encountered in the past, but there is still no substitute for the integrity and judgment of the good workman. It must be confessed however, that some miserably low-grade lumber is available. While it is entirely possible that at times, in some locations, satisfactory results may be attained when using low-grade lumber, this lumber must be used with caution. While it is entirely possible that, at times in some locations, satisfactory results may be attained when using low-grade lumber. This lumber must be used with caution. There is little room for discrimination between lumber species. Any of the species found in the stock of the average lumber yard can be used for most purposes.

The grades under different grading rules are confusing, but in general, the following list gives approximate equivalents:

No. 1 kiln-dried yellow pine dimension is approximately equivalent to construction-grade Douglas fir joists and planks.

No. 2 yellow pine dimension is approximately equivalent to Standard-grade, and superior to structural-grade Douglas fir.

No. 1, 2, and 3 Douglas fir dimension, utility-grade Douglas fir, and No. 3 yellow pine are not stress-graded, quality is quite low, and only the judgment of the user can govern their uses.

Among woods of other species, dense structural redwood, No. 1 dense yellow pine dimension, select structural western hemlock light framing, select structural western larch joists and plank, dense select structural Douglas fir joists and plank, and structural cypress joists and plank are all fine premium-grade timbers, but, of course, at premium prices. Their use is not often warranted in ordinary construction.

A knot is a defect that may or may not reduce the strength of the timber. A knot is formed from the growth of a limb or branch. As the limb is surrounded by new growth the grain is bent. As long as the limb grows, new layers of growth will be added to the limb. This type of knot is called an intergrown knot. An encased knot is formed when a limb dies and the trunk continues to grow, placing growth layers around the knot.

Season checks in timber may or may not be a source of slight weakness. They are more injurious on the vertical than on the horizontal face of a stringer or joist, and their effect continues even after they close up, as many do, and are no longer visible; such defects can only be found by careful examination.

For light framing, the generally-obtainable commercial softwoods, especially Southern yellow pine and Douglas fir, are generally to be preferred, unless the lower price of locally obtained timbers of other species make them more attractive. Only a few of the hardwoods have strength and stiffness comparable with longleaf yellow pine. In some cases, where attractive appearance is desired, the hardwoods may be preferred, as in floors and interior trim. Where resistance to abrasion is desirable, hard-maple flooring is unexcelled.

Many different species of timber may be used for the same purposes, in most cases. The following are quite desirable for interior trim that is to be varnished: cypress, western red cedar, sap shortleaf yellow pine, and almost any of the hardwoods. White pine or ponderosa pine are desirable for enameling. For outside trim, cypress and redwood are our two most-durable timbers, but unfortunately neither of them take paint exceptionally well, and redwood seasons exceedingly slow. Southern yellow pine is somewhat rot resistant and is used extensively in cornice construction.

Ordering Lumber

Practically all yard lumber is now furnished in uniform widths, lengths in even feet from 4 to 20 feet, with the exception of end-matched flooring and D & M, which comes in random lengths. The edges may be shiplapped, dressed and matched (D & M), or square. With standard crib siding, the edges are beveled. All thicknesses less than 1 inch are counted as 1 inch thick. Most, but not all, widths are in multiples of 2 inches. Unless specified, widths and thicknesses are only nominal—the sizes of the rough lumber from which the boards were worked.

Properties of Woods

Certain species of timber are more desirable than others for given jobs, and it is convenient to know the properties and physical characteristics of various woods. Following is a partial list of species which are generally obtainable.

White or Gray Ash—Hard, heavy, and springy; light reddish-brown heart; sapwood nearly white. Too hard to nail when dry.

Several species are marketed together as white ash. The woods are practically identical.

Brown Ash—Not a framing timber, but an attractive trim wood. Brown heart, lighter sapwood. The trees often wind shake so badly that the heart is entirely loose. Attractive veneers are sliced from stumps and forks.

White Cedar, Northern—Light-brown heart, sapwood thin and nearly white. Light, weak, soft, decay resistant, holds paint well. Marketed with Atlantic white cedar.

Western Red Cedar—Also called canoe cedar or shinglewood. Light, soft, straight-grained, small shrinkage, holds paint well. Heart is light brown, extremely rot resistant. Sap quite narrow, nearly white. Used for shingles, siding, boat building.

Eastern Red Cedar, or Juniper—Pungent aromatic odor said to repel moths. Red or brown heartwood, extremely rot resistant white sapwood. Used for lining clothes closets and chests and for fence posts.

Cypress—Probably our most durable wood for contact with the soil. Wood moderately light, close-grained, heartwood red to nearly yellow, sapwood nearly white. Does not hold paint well, but otherwise desirable for siding and outside trim. Attractive for inside trim.

Gum, Red—Moderately heavy, interlocking grain; warps badly in seasoning; heart is reddish brown, sapwood nearly white. The sapwood may be graded out and sold as white gum, the heartwood as red gum, or together as unselected gum. Cuts into attractive veneers.

Hickory—A combination of hardness, weight, toughness, and strength found in no other native wood. A specialty wood, almost impossible to nail when dry. Not rot resistant. Several related species marketed together.

Hemlock, Eastern—Heartwood is pale brown to reddish, sapwood not distinguishable from heart. May be badly wind shaken. Brittle, moderately weak, not at all durable. Used for cheap, rough framing veneers.

Hemlock, Western—Heartwood and sapwood almost white with purplish tinge. Moderately strong, not durable, mostly used for pulpwood. Some marketed mixed with Douglas fir.

Locust, Black—Heavy, hard, strong; the heartwood is exceptionally durable. Not a framing timber. Used mostly for posts and poles.

Maple, Hard—Heavy, strong, hard, and close grained; color light brown to yellowish. Used mostly for wear-resistant floors, and furniture. Circularly growing fibers cause the attractive "birds-eye" grain in some trees. One species, the Oregon maple, occasionally contains the attractive "quilted" grain.

Maple, Soft—Softer and lighter than hard maple; lighter colored. Box elder is sometimes marketed with soft maple. Used for much the same purposes as hard maple, but not nearly so desirable.

Oak, White—Several species are marketed together, but the woods are practically identical. Hard, heavy, tough, strong, and somewhat rot-resistant. Brownish heart, lighter sapwood. Desirable for trim and flooring, and one of our best hardwood framing timbers.

Oak, Red—Several species are marketed together. They cannot be distinguished one from the other, but can be distinguished from the white oaks. Good framing timber, but not rot resistant.

Pine, White, Western—Also called Idaho white pine. Creamy or light-brown sapwood, sapwood thick and white. Used mostly for millwork and siding. Moderately light, moderately strong, easy to work, holds paint well.

Pine, Red or Norway—Resembles the lighter weight specimens of Southern yellow pine. Moderately strong and stiff, moderately soft, heartwood pale red to reddish brown. Used for millwork, siding, framing, and ladder rails.

Pine, Long-Leaf Southern Yellow—Not a species but a grade. All Southern yellow pine that has six or more annual rings per inch is marketed as long-leaf, and it may contain lumber from any of the several species of Southern pine. Heavy, hard, and strong, but not especially durable in contact with the soil. The sapwood takes creosote well. One of our most useful timbers for light framing.

Pine, Short-Leaf Southern Yellow—Contains timber from any of several related species of Southern pine having less than six annual rings per inch. Quite satisfactory for light framing, and the sapwood is attractive as an interior finish.

21

Douglas Fir—Our most plentiful commercial timber. Varies greatly in weight, color, and strength. Strong, moderately heavy, splintery, splits easily. Used in all kinds of construction; much is rotary cut for plywood.

Poplar, Yellow or Tulip—Our easiest-working native wood. Old growth has a yellow to brown heart. Sapwood and young trees are tough and white. Not a framing lumber, but used to a great extent for siding, where it may be marketed with cucumber magnolia, a botanical relative.

Redwood—One of our most durable and rot-resistant timbers. Light, soft, moderately high strength, heartwood reddish brown, sapwood white. Does not paint exceptionally well, as it oftentimes "bleeds" through. Used mostly for siding and outside trim.

Spruce, Sitka—Light, soft, medium strong, heart is light reddish brown, sapwood is nearly white, shading into the heartwood. Usually cut into boards, planing-mill stock, and boat lumber.

Spruce, Eastern—Stiff, strong, hard, and tough. Moderately light weight, light color, little difference between heart and sapwood. Commercial eastern spruce includes wood from three related species. Used for pulpwood, framing lumber, millwork, etc.

Spruce, Engelmann—Color, white with red tint. Straight grained, light weight, low strength. Used for dimension lumber and boards, and for pulpwood. Extremely low rot resistance.

Tamarack, or Larch—Small to medium sized trees; not much is sawed into framing lumber, but much is cut into boards. Yellowish brown heart, sapwood white. Much is cut into posts and poles.

Walnut, Black—Our most attractive cabinet wood. Heavy, hard, and strong, heartwood is a beautiful brown, sap nearly white. Mostly used for fine furniture, but some is used for fine interior trim. Somewhat rot-resistant. Used also for gunstocks.

Walnut, White or Butternut—Sapwood light to brown, heart light chestnut brown with an attractive sheen. The cut is small, mostly going into cabinet work and interior trim. Moderately light, rather weak, not rot-resistant.

Wood Preservatives

By using certain chemical wood preservatives, timbers which are not at all rot-resistant can be used under conditions that are favorable to rot. Coal-tar creosote is the preservative most widely used, and probably it is the most effective, but there are many other chemicals that are more or less effective. Among them are: pentachlorophenol, probably next to creosote as the most popular; and the patented preparations Tanalith, Celcure, ZMA, Chemonite, Greensalt, Boliden Salt, and others. Most timber treatment is done by commercial firms, using one of the pressure processes. For some of the water-borne preservatives, the wood should be entirely green, but for most of them, the wood should be air-dried for three to six months before treatment.

Brush coats of preservatives are generally of little value, but thorough soaking for three to six weeks is normally quite effective. If possible, all cutting and framing should be done before treatment. All of the preservative treatments are effective against termites.

The odor of creosote is distinctive, penetrating, and unpleasant to some persons, and it may burn the skin of some, something like a sunburn, but creosote is never a hazard to life or health. Some judgment should be exercised when using creosoted framing lumber, especially where flooring or other lumber finishes must be nailed directly to the creosoted timber. The oils follow the nails, and sometimes a spot may appear around each nail. It is difficult or practically impossible to paint over creosote.

Creosoted timber, when freshly treated, does ignite rather readily, and it burns freely. Where there is a serious fire hazard, some less volatile preservative should be used.

Decay of Timber

Decay of timber is the result of one cause, and one cause only, the work of certain low order plants called fungi. All of these organisms require water and air to live, grow, and multiply; consequently, wood that is kept dry or that is dried quickly after wetting, will not decay. Similarly, wood that is kept submerged in water will soften, but it will not decay, for the air supply is shut off, and timber set deep in the ground, such as piling, which is shut off from the air, is practically permanent.

There is no such thing as "dry rot"; however, the term is rather loosely used sometimes when speaking of about any type of dote,

23

or any punky, dry, and decayed wood. Although such rotten wood may be dry when observed, it was wet while decay was progressing. This kind of decay is often found inside living, growing trees, but it only occurs in the presence of water.

MODIFIED FORMS OF WOOD

Lumber and timbers are, of course, the most important products of our forests, but aside from these, many other products have appeared within relatively recent years and are of great interest to the builder. Some of these may be called modifications of wood; many of them are adaptions. The second most important product of the forests is pulpwood; the pulp for the most part is utilized in making paper, but much is also used to make various types of building boards. Various chemicals are produced, mostly from the wastes of the paper industry, including sulfate turpentine, used in the paint and dye industries; from tall oil, a waste of kraft paper manufacture, comes fatty acids and resins that are used in making soaps, detergents, adhesives, and insecticides.

The development of plywood has given us large, thin sheets, a modification of wood without the inherent defects of the natural product. The development of glue-laminated beams and arches is another modification, or rather adaptation, of this material. The development of wood-pulp building boards, hardboards, and softboards is another modification. All of these products were undreamed of when our pioneers first began using timber to build shelters for themselves.

Laminated Wood

In days gone by, the use of wood as a building material was confined to relatively small pieces. These light and easily handled members served their purpose admirably for almost all types of residential construction. Timbers were occasionally used in some comparatively heavy mill and shop construction, but the use of timbers was governed by the sizes and quality of the timbers available, which were limited by the sizes of the trees, by the defects inherent in all solid timbers, by difficulties in drying, and some other factors. Since the advent of glue-laminated timbers, these difficulties have been largely overcome. The laminations are relatively small, thin pieces, and they can be selected so that they are free from damaging imperfections. Since they are thin, no difficulty is encountered in drying. Casein glues are generally used; these are water-resistant and gap filling. The thin laminations can be readily bent to almost any form, and hence the laminated arch of wood has been developed. Wood is now taking its rightful place as an engineering material among other materials such as steel and reinforced concrete. Many, perhaps most, churches and auditoriums utilize glue-laminated wood arches or beams. Wood beams have been built with spans of more than 100 feet, and wood arches have been built for clear spans of more than 200 feet.

Plywood

The most familiar type of plywood used in the United States is made from Douglas fir. Short logs are chucked into a lathe, and a thin, continuous layer of wood is peeled off. This thin layer is straightened, cut to convenient sizes, covered with glue, laid up with the grain in successive plies crossing, and subjected to heat and pressure. This is the plywood of commerce, one of our most useful building materials.

All plywood has an odd number of plies, allowing the face plies to have parallel grain while the lay-up is "balanced" on each side of a center ply. This process equalizes stresses set up when the board dries or when it is subsequently wetted and dried.

In general, the thickness of the face plies is ⅛ inch, but under some circumstances, using certain species of wood, thinner plies are advisable. An exceptionally high quality birch plywood, 1 inch in thickness and with 19 plies, is imported from Finland. Our own standard Douglas-fir plywood consists of three plies of ¼, $\frac{5}{16}$, and ⅜ inch, five plies for ½, ⅝, and ¾ inch, and seven plies for ⅞ to $1\frac{3}{16}$ inches. Many modifications are available; among these are "overlaid" panels with resin-impregnated faces, textured and grooved patterns of various types, and embossed or striated faces. Plywood has also been produced with a heat-insulating foamed styrene core. Finish plywoods with face plies consisting of all our finest native hardwoods and many exotic species are also available.

Wood-Product Board Materials

Various types of building boards, basically wood products, have an important use in housing. There are three general types—the fiberboards, weighing 15 to 25 pounds per cubic foot, the particle boards, weighing 26 to 50 pounds per cubic foot, and the hardboards, weighing 51 to 75 pounds per cubic foot. Each is used for a specialized purpose. The fiberboards, or softboards, are used primarily to provide heat insulation, since they have little or no structural strength. The particle boards have considerable strength, hold screws and nails reasonably well, and are used for cabinet work and shelving. The hardboards are used as underlays for flooring materials and for cabinet sides and backs; they are produced in the form of siding for exterior use, since they hold paints and other finishes reasonably well. Besides these rigid and semirigid boards, wood fibers are made into flexible mats and fill-type insulations.

Semirigid Boards

Semirigid insulating boards are made from other materials than wood pulp, including bagasse, or cane-mill refuse, and flax straw. They are all manufactured in essentially the same way. The water-suspended pulp is run through a paper processing machine, or similar machine, and is reduced to a wet mat. The water is then squeezed out, and the mat is pressed to the desired thickness and dried. Bonding agents are sometimes used, but rarely in significant quantity; the boards depend on simple felting of the fibers to provide rigidity. Most of the softboards have considerable value as acoustical materials, because they are good absorbers of sound.

Particle Boards

In the manufacture of these boards, the structure of the wood is not broken down, but simply reduced to flakes, or particles, which are bound together with a synthetic resin, often urea-formaldehyde or phenol-formaldehyde. The boards are then cured under heat and moderate pressure.

Hardboards

Hardboard is made from wood pulp, and it usually contains no binder other than the slightly thermoplastic lignin in the wood. The board is formed under heat and heavy pressure. One side of the board is smooth; the other may or may not have a rough screen surface. It may be tempered by impregnating the board with oil and then baking it.

Other Fibrous Materials

Flexible insulating materials are made from many substances, including glass, rock, and mineral wools, wood fibers, and other such materials. They may or may not have a covering of paper, and they are normally called blankets or batts. Fill insulation is made from rock, glass, or mineral wool, perlite (a burned volcanic residue), vermiculite (or fused mica), and shredded redwood bark.

SUMMARY

There are many basic methods of preparing lumber for the market. It is necessary that it be seasoned (moisture removed), this process being classed as natural and artificial. Natural seasoning consists of exposing sawed lumber to free circulating air. Artificial drying, or kiln drying, is accomplished in most cases by forcing heated air over the lumber, thereby removing the moisture. When the moisture content of lumber is down to 12 to 19 percent, the lumber is said to be air-dry.

The selection of various grades of lumber to be used for any specific purpose is often left to the discretion of the carpenter. Defects often found in manufactured lumber are shakes, knots, pitch-pockets, imperfect machining, machine burns, and splits from imperfect drying. Knots, coarse grain, and other defects may or may not reduce the strength of the lumber, depending on their location in the piece.

By using certain wood preservative chemicals, timbers which are not rot-resistant can be used under conditions that are favorable to rot. Preservatives most widely used are coal-tar creosote, pentachlorophenol, Celcure, Chemonite, Greensalt, and Boliden Salt. Most timber treatment is done by commercial firms, using one of the pressure processes.

Many other products have appeared in recent years from the lumber and timber industry. Pulpwood is just one product which

27

is utilized in making paper, but much is also used to make various types of building boards. Various chemicals are produced, mostly from the paper industries, including sulfate turpentine used in paint and dye products, and fatty acids and resins used in producing soaps, detergents, and adhesives. The development of plywood and glue-laminated beams and arches is one of our most useful building materials.

REVIEW QUESTIONS

1. Define the word logging.
2. How may lumber be seasoned? Explain.
3. What are some of the defects found in lumber?
4. What should a person look for when purchasing framing lumber?
5. Name a few products derived from the lumber and timber industry.

CHAPTER 2

Nails

Up to the end of the Colonial period, all nails used in the United States were handmade. They were forged on an anvil from nail rods, which were sold in bundles. These nail rods were prepared either by rolling iron into small bars of the required thickness, or by the much more common practice of cutting plate iron into strips by means of rolling shears.

Just before the Revolutionary War, the making of nails from these rods was a household industry among the New England farmers. The struggle of the Colonies for independence intensified an inventive search for short cuts to mass production of material entering directly or indirectly into the prosecution of the war; thus came about the innovation of cut nails made by machinery. With its coming, the household industry of nail making rapidly declined. At the close of the 18th century, 23 patents for nail-making machines had been granted in the United States, and their use had been generally introduced into England, where they were received with enthusiasm.

In France, light-weight nails for carpenter's use were made of wire as early as the days of Napoleon I, but these nails were made by hand with a hammer. The handmade nail was pinched in a vise with a portion projecting. A few blows of a hammer flattened one end into a head. The head was beaten into a countersunk depression in the vise, thus regulating its size and shape. In the United States, wire nails were first made in 1851 or 1852 by William Hersel of New York.

In 1875, Father Goebel, a Catholic priest, arrived from Germany and settled in Covington, Ky.; there he began the manufacture of wire nails which he had learned in his native land. In 1876, the American Wire and Screw Nail Company was formed under Father Goebel's leadership. As the production and consumption of wire nails increased, the vogue of cut nails, which dominated the market until 1886, declined.

The approved process of the earlier days of the cut-nail industry was as follows: Iron bars, rolled from hematite or magnetic pig were fagotted, reheated to a white heat, drawn, rolled into sheets of the required width and thickness, and then allowed to cool. The sheet was then cut across its length (its width being usually about a foot) into strips a little wider than the length of the required nail. These plates, heated by being set on their edge on hot coals, were seized in a clamp and fed to the machine, end first. The cut out pieces, slightly tapering, were squeezed and headed up by the machine before going to the trough.

The manufacture of tacks, frequently combined with that of nails, is a distinct branch of the nail industry, affording much room for specialties. Originally it was also a household industry carried on in New England well into the 18th century. The wire, pointed on a small anvil, was placed in a pedal-operated vise, which clutched it between jaws furnished with a gauge to regulate the length. A certain portion was left projecting; this portion was beaten with a hammer into a flat head.

Antique pieces of furniture are frequently held together with iron nails that are driven in and countersunk, thus holding quite firmly. These old-time nails were made of four-square wrought iron and tapered somewhat like a brad but with a head which, when driven in, held with great firmness.

The raw material of the modern wire nail factory is drawn wire, just as it comes from the wire-drawing block. The stock is low-carbon Bessemer or basic open-hearth steel. The wire, feeding from a loose reel, passes between straightening rolls into the gripping dies, where it is gripped a short distance from its end, and the nail head is formed by an upsetting blow from a heading tool. As the header withdraws, the gripping dies loosen, and the straightener carriage pushes the wire forward by an amount equal to the length of the nail. The cutting dies advance from the sides of the frame and clip off the nail, at the same time forming its characteristic chisel point. The gripping dies have already seized the wire again, and an ejector flips the nail out of the way just as the header comes forward and heads the next nail. All these motions are induced by cams and eccentrics on the main shaft of the machine, and the speed of production is at a rate of 150 to 500 or more complete cycles per minute. At this stage, the nails are covered with a film of drawing lubricant and oil from the nail machine, and their points are frequently adorned with whiskers—a name applied to the small diamond-shaped pieces stamped out when the point is formed and which are occasionally found on the finished nail by the customer.

These oily nails, in lots of 500 to 5000 pounds, are shaken with sawdust in tumbling barrels from which they emerge bright and clean and free of their whiskers, ready for weighing, packing, and shipping.

THE "PENNY" SYSTEM

This method of designating nails originated in England. Two explanations are offered as to how this curious designation came about. One is that the six penny, four penny, ten penny, etc., nails derived their names from the fact that one hundred nails cost six pence, four pence, etc. The other explanation, which is the more probable of the two, is that one thousand ten-penny nails, for instance, weighed ten pounds. The ancient, as well as the modern, abbreviation for penny is *d,* being the first letter of the Roman coin denarius; the same abbreviation in early history was used for the English pound in weight. The word *penny* has persisted as a term in the nail industry.

KINDS OF NAILS

Nails are the carpenter's most useful fastener, and a great variety of types and sizes are available to meet the demands of the industry. One manufacturer claims to produce more than 10,000 types and sizes. Some common types of nails are illustrated in Fig. 1.

The following shapes of points are available:
1. Common blunt pyramidal.
2. Long sharp.
3. Chisel-shaped.
4. Blunt, or shooker.
5. Side-sloped.
6. Duck-bill, or clincher.

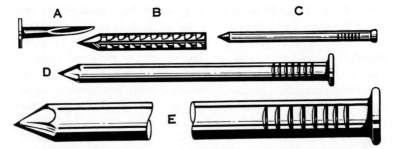

Fig. 1. Various nails grouped as to general size: A, tack; B, sprig or dowel pin; C, brad; D, nail; E, spike.

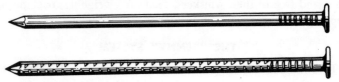

Fig. 2. Smooth and barbed box nails, 16d size (shown full size). Note the sharp point and thin, flat head.

The shanks, as shown in Fig. 2, may be:
1. Barbed.
2. Longitudinally grooved.
3. Round smooth.
4. Spiral grooved.
5. Annular grooved.

Nails may be finished:
1. Bright.
2. Galvanized, usually hot-dipped.
3. Cadmium-plated.
4. Blued.
5. Painted.
6. Cement-coated.

Nails may be made of:
1. Mild steel.
2. Copper.
3. Brass.
4. Aluminum.
5. Case-hardened steel.

32

The heads may be:
1. Flat.
2. Oval or oval countersunk.
3. Round.
4. Countersunk.
5. Double-headed.
6. Cupped.
7. Lead-headed.

Besides these, there are many nails with other types of heads adapted for special uses.

Tacks

Tacks are small, sharp pointed nails that usually have tapering sides and a thin, flat head. The regular lengths of tacks range from ⅛ to 1⅛ inches. The regular sizes are designated in ounces, according to Table 1.

Table 1. Wire Tacks

Size oz.	Length in.	No. per pound	Size oz.	Length in.	No. per pound	Size oz.	Length in.	No. per pound
1	1/8	16,000	4	7/16	4000	14	13/16	1143
1 1/2	3/16	10,666	6	9/16	2666	16	7/8	1000
2	1/4	8000	8	5/8	2000	18	15/16	888
2 1/2	5/16	6400	10	11/16	1600	20	1	800
3	3/8	5333	12	3/4	1333	22	1 1/16	727
....	....	....	....	....	....	24	1 1/8	666

Sprigs

The name *sprig* is sometimes given to a small headless nail, usually called a barbed dowel pin. Sprigs are made regularly in sizes ½ to 2 inches; No. 8 steel wire gauge or 0.162 inch diameter.

Brads

Brads are small slender nails with small deep heads; sometimes, instead of having a head, they have a projection on one side. There are several varieties adapted to many different requirements. Although brads are generally thought of as being very small, the common variety is made in sizes from ¼ inch.

Table 2. Common Nails

| | Plain | | | Coated | | | |
Size	Length in.	Gauge No.	No. per pound	Length in.	Gauge No.	No. per Keg	Net Wgt. pounds
2d	1	15	876	1	16	85,700	79
3d	1¼	14	568	1⅛	15½	54,300	64
4d	1½	12½	316	1⅜	14	29,800	61
5d	1¾	12½	271	1⅝	13½	25,500	70
6d	2	11½	181	1⅞	13	17,900	65
7d	2¼	11½	161	2⅛	12½	15,300	72
8d	2½	10¼	106	2⅜	11½	10,100	71
9d	2¾	10¼	96	2⅝	11½	8900	68
10d	3	9	69	2⅞	11	6600	63
12d	3¼	9	63	3⅛	10	6200	80
16d	3½	8	49	3¼	9	4900	80
20d	4	6	31	3¾	7	3100	83
30d	4½	5	24	4¼	6	2400	84
40d	5	4	18	4¾	5	1800	82
50d	5½	3	14	5¼	4	1300	79
60d	6	2	11	5¾	3	1100	82

Nails

The term "nails" is popularly applied to all kinds of nails except extreme sizes, such as tacks, brads, and spikes. Broadly speaking, however, it includes all of these. The most generally used are called common nails, and are regularly made in sizes from 1 inch (2d) to 6 inch (60d), as shown in Table 2 and Figs. 3, 4, and 5. Some special types of nails are illustrated in Figs. 6 through 10.

Spikes

By definition, an ordinary spike is a stout piece of metal from 3 to 12 inches in length and thicker in proportion than a common nail. It is provided with a head and a point and is frequently curved, serrated, or cleft to render extraction difficult. It is used to a great extent to attach railroad rails to ties and is also used in the construction of docks, piers, and other work requiring large timbers.

It should be noted that spike and common-nail sizes overlap; sizes common to both are from 3 to 6 inches, the spike being thicker for equal sizes. There are two kinds of ordinary or round wire spikes classed with respect to the shape of the ends, as flat heads, diamond point and oval head, chisel point. The sizes and other proportions for ordinary spikes are given in Table 3.

HOLDING POWER OF NAILS

Numerous tests have been made at various times to determine the holding power of nails. Tests at the Watertown Arsenal on different sizes of nails from 8*d* to 60*d* gave average results in pounds, as shown in Table 4.

A. M. Wellington found the force required to withdraw spikes ⁹⁄₁₆ × ⁹⁄₁₆ inch, driven 4¼ inches into seasoned oak, to be 4281 pounds; the same spikes driven into unseasoned oak, 6523 pounds.

Professor W. R. Johnson found that a plain spike ⅜ inch square, driven 3⅜ inches into seasoned yellow pine or unseasoned chestnut required approximately 2000 pounds of force to extract it; from seasoned white oak, approximately 4000 pounds; and from well-seasoned locust, 6000 pounds.

Experiments in Germany, by Funk, give from 2465 to 3940 pounds (the mean of many experiments was 3000 pounds) as the force necessary to extract a plain ½-inch square iron spike 6 inches long, wedge-pointed for 1 inch and driven 4½ inches into white or yellow pine. When driven 5 inches, the force required was approximately 1/10 part greater. Similar spikes 9/16 inches square, 7 inches long, driven 6 inches deep, required from 3700 to 6745 pounds of force to extract them from pine; the mean of the results was 4873 pounds. In all cases, about twice as much force was required to extract them from oak. The spikes were all

Table 3. Ordinary Spikes

Size	Length in.	Gauge No.	Degree of Countersink	Head Diam.	Head Rad.	No. per pound
10*d*	3	6	123	13/32	7/16	41
12*d*	3 1/4	6	....	....	....	38
16*d*	3 1/2	5	123	7/16	7/16	30
20*d*.	4	4	123	15/32	7/16	23
30*d*	4 1/2	3	123	1/2	7/16	17
40*d*	5	2	123	17/32	7/16	13
50*d*	5 1/2	1	....	....	....	10
60*d*	6	1	123	9/16	7/16	9
7 inch	7	5/16 inch	123	5/8	5/8	7
8 "	8	3/8 "	123	3/4	3/4	4
9 "	9	3/8 "	....	....	....	3 1/2
10 "	10	3/8 "	....	....	....	3
12 "	12	3/8 "	....	....	....	2 1/2

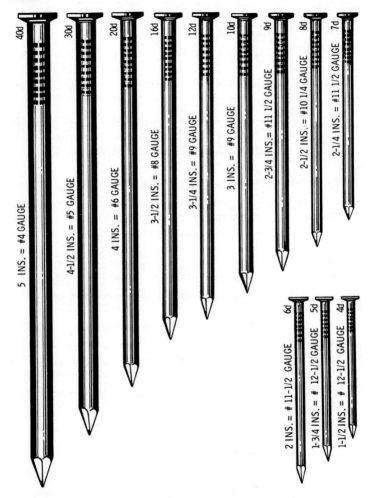

40d

30d

20d

16d

12d

10d

9d

8d

7d

5 INS. = #4 GAUGE

4-1/2 INS. = #5 GAUGE

4 INS. = #6 GAUGE

3-1/2 INS. = #8 GAUGE

3-1/4 INS. = #9 GAUGE

3 INS. = #9 GAUGE

2-3/4 INS. = #11 1/2 GAUGE

2-1/2 INS. = #10 1/4 GAUGE

2-1/4 INS. = #11 1/2 GAUGE

6d

5d

4d

2 INS. = # 11-1/2 GAUGE

1-3/4 INS. = # 12-1/2 GAUGE

1-1/2 INS. = # 12-1/2 GAUGE

Fig. 3. Common wire nails—the standard nail for general use is regularly made in sizes from 1 inch (2d) to 6 inches (60d).

driven across the grain of the wood. When driven with the grain, spikes or nails do not hold with more than half as much force.

Boards of oak or pine nailed together by from 4 to 16 ten-penny common, cut nails and then pulled apart in a direction lengthwise to the boards, and across the nails (tending to break the latter in two by a shearing action) averaged 300 to 400 pounds per nail

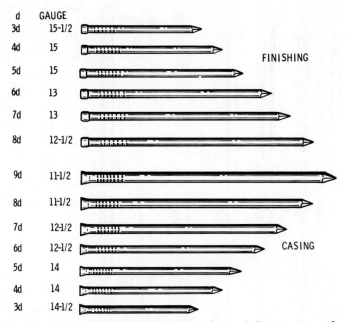

d	GAUGE		
3d	15-1/2		
4d	15		FINISHING
5d	15		
6d	13		
7d	13		
8d	12-1/2		
9d	11-1/2		
8d	11-1/2		
7d	12-1/2		
6d	12-1/2		CASING
5d	14		
4d	14		
3d	14-1/2		

Fig. 4. Various casing and finishing nails (shown full size). Note the difference in the shape of heads and size of wire. The finishing nail is larger than a casing nail of equal length.

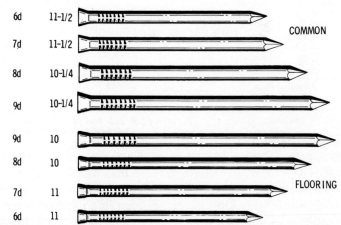

6d	11-1/2		COMMON
7d	11-1/2		
8d	10-1/4		
9d	10-1/4		
9d	10		
8d	10		
7d	11		FLOORING
6d	11		

Fig. 5. Flooring and common nails (shown full size). Note the variation in head shape and gauge number.

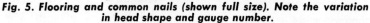

37

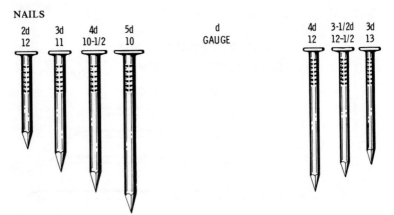

NAILS

2d	3d	4d	5d
12	11	10-1/2	10

d
GAUGE

4d	3-1/2d	3d
12	12-1/2	13

Fig. 6. *A few sizes of slating and shingle nails. Note the difference in wire gauge.*

Fig. 7. *Hook-head, metal-lath nail. This is a bright, smooth nail with a long, thin, flat head, made for application of metal lath. It is also made blued or galvanized.*

Table 4. Withdrawal Force
(lbs. per sq. in. of surface)

Wood	Wire Nail	Cut Nail
White Pine	167	405
Yellow Pine	318	662
White Oak	940	1216
Chestnut		683
Laurel	651	1200

to separate them. Chestnut offers about the same resistance as yellow pine.

A. W. Wright of the Western Society of Engineers obtained the following results with spikes driven into dry cedar (cut 18 months):

With respect to types of shanks, the plain-shank, low-carbon, common nails have sufficient holding power for most work. The

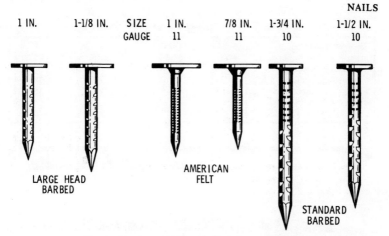

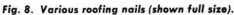

Fig. 8. Various roofing nails (shown full size).

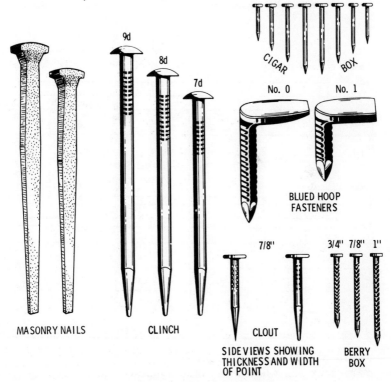

Fig. 9. Miscellaneous nails.

39

Table 5. Holding Power of Spikes

Size of spikes	5×¼ in. sq.	6×¼	6×½	5×⅜
Length driven in	4¼ in.	5 in.	5 in.	4¼ in.
Pounds resistance to				
drawing, average lbs.	857	857	1691	1202
max. lbs.	1159	923	2129	1556
From 6 to 9 tests each				
min. lbs.	766	766	1120	687

Fig. 10. Hinge nails.

Table 6. Holding Power of Nails and Spikes (Withdrawal)

	Cut Nails					
	Parallel to grain			Cross grain		
Size	Yellow Pine	White Pine	White Oak	Yellow Pine	White Pine	White Oak
6d	89	......	......	154	77	317
8d	206	89	520	327	211	630
10d	222	108	580	324	181	650
20d	320	148	692	407	298	800
50d	439	170	820	570	316	991
60d	445	200	950	639	324	1040

	Wire Nails				
	Parallel to grain	Cross grain			
Size	White Pine	Cedar (dry)	White Oak	Yellow Pine	White Pine
6d	30	......	129	108	60
10d	50	......	390	132	70
60d	......	......	731	465	......
3/8 in.	370	283	1,188	590	450
1/16 in.	344	......	......	......	436
1/2 in.	113	338	744	700	364

barbed nail does not have quite as much holding power as the plain-shank nail when driven into dry wood, but driven into wet or green wood, they do not lose their grip when the wood dries out as do nails with plain shanks. Coated nails are often used for short-time holding power, such as for boxing and crating. In the smaller sizes, the holding power of cement-coated nails may be as much as 150% greater than that of plain uncoated nails, especially when driven into soft wood, although in very hard woods, the holding power may be little, if any, greater. In any case, the extra holding power is lost in a relatively short time.

The increase in holding power of deformed-shank and cement-coated nails is for withdrawal resistance only. There is little or no difference in the shearing resistance of all nails with the same sizes of shank. In buildings, nails are always placed in shear if at all possible, and the extra cost of deformed or coated nails may or may not be justified.

SELECTION OF NAILS

On any kind of construction work, an important consideration is the type and size of nails to use. The first factor is the finish. Should the nails be smooth, barbed, or cement coated? The holding power of cement-coated nails, it was found, is considerably greater than that of the same sized smooth nails. In most cases, the barbed nails have the least holding power. Thus, nails can be graded with regard to holding power as follows: first, cement-coated; second, smooth; third, barbed.

Next to be considered is the diameter size of the nail. Short, thick nails work loose quickly. Long, thin nails are apt to break

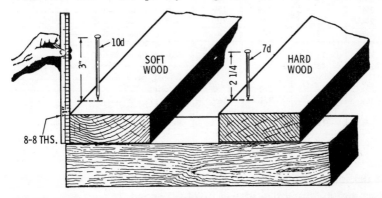

Fig. 11. Application of rules 2 and 3 in determining the proper size of nail to use.

41

Table 7. Wire Nails—Kinds and
(American Steel

Length, in inches	Am. Steel & Wire Co.'s Steel Wire Gauge	Approx. No. to lbs.	Nailings	Sizes and Kinds of Material		
2½	10¼	106	2	1× 4		
2½	10¼	106	2	1× 6		
2½	10¼	106	2	1× 8		
2½	10¼	106	2	1×10		
2½	10¼	106	3	1×12		
4	6	31	2	2× 4	Used square edge, as platforms, floors, sheathing, or shiplap.	When used D. & M., blind nailed, only ½ quantity named required.
4	6	31	2	2× 6		
4	6	31	2	2× 8		
4	6	31	3	2×10		
4	6	31	3	2×12		
6	2	11	3	2× 4		
6	2	11	2	3× 6		
6	2	11	2	3× 8		
6	2	11	3	3×10		
6	2	11	3	3×12		
2½	12½	189	2	Base, per 100 ft. lin.		
2½	10¼	106	2	Byrket lath		
2½	12½	189	1	Ceiling, ¾×4		
2	13	309	1	Ceiling, ½ and ⅝		
2½	12½	189	2	Finish, ⅞		
3	11½	121	2	Finish, 1⅛		
2½	10	99	1	Flooring, 1×3		
2½	10	99	1	Flooring, 1×4		
2½	10	99	1	Flooring, 1×6		
4	6	31		Framing, 2×4 to 2×16 requires		
3½	8	49		9 or more sizes and vary		
3	9	69		greatly.		
6	2	11		Framing, 3×4 to 3×14		
2½	11½	145	2	Siding, drop, 1×4		
2½	11½	145	2	Siding, drop 1×6		
2½	11½	145	2	Siding, drop 1×8		
2	13	309	1	Siding, bevel, ½×4		
2	13	309	1	Siding, bevel, ½×6		
2	13	309	1	Siding, bevel, ½×8		
				Casing, per opening		
1¼	14	568	12″ o. c.	Flooring, ⅜×2		
1½	15	778	16″ o. c.	Lath, 48″		
⅞	12	469	2″ o. c.	Ready roofing		
⅞	12	469	1″ o. c.	Ready roofing		

at the joints of the lumber. The simple rule to follow is to use as long and as thin a nail as will drive easily.

Definite rules have been formulated by which to determine the

Quantities Required (continued)

and Wire Co.)

Trade Names	Pounds per 1000 feet B. M. on center as follows:				
	12"	16"	20"	36"	48"
	Pounds				
8d common	60	48	37	23	20
8d common	40	32	25	16	18
8d common	31	27	20	12	10
8d common	25	20	16	10	8
8d common	31	24	20	12	10
20d common	105	80	65	60	32
20d common	70	54	43	27	22
20d common	53	40	53	21	17
20d common	60	50	40	25	20
20d common	52	41	33	21	17
60d common	197	150	122	76	61
60d common	131	97	82	52	42
60d common	100	76	61	38	34
60d common	178	137	110	70	55
60d common	145	115	92	53	46
8d finish	....	1			
8d common	....	48			
8d finish	18	14			
6d finish	11	8			
8d finish	25	12			
10d finish	12	10			
8d flooring	42	32			
8d flooring	32	26			
8d flooring	22	18			
⌈ 20d common	20	16	14		
| 16d common	10	10	8		
| 10d common	8	6	5		
⌊ 60d common	30	25	20		
8d casing	45	35			
8d casing	30	25			
8d casing	23	18			
6d finish	23	18			
6d finish	15	13			
6d finish	12	10			
6d and 8d casing.	About ½ pound per side.				
3d brads	About 10 pounds per 1000 square feet.				
3d fine	6 pounds per 1000 pieces.				
Barbed roofing	¾ of a pound to the square.				
Barbed roofing	1½ pounds to the square.				

size of nail to be used in porportion to the thickness of the board that is to be nailed:

1. When using box nails in timber of medium hardness, the

Table 7.   Wire Nails—Kinds and
(American Steel

Length, in inches	Am. Steel & Wire Co.'s Steel Wire Gauge	Approx. No. to lbs.	Nailings	Material Sizes and Kinds of
⅞	12	180	2" o. c.	Ready roofing .... (⅝ heads)
⅞	12	180	1" o. c.	Ready roofing .... (⅝ heads)
1¼	13	420	......	Shingles† ....
1½	12	274	......	Shingles ....
⅞	12	180	4	Shingles ....
⅞	12	469	4	Shingles ....
1	16	1150	2" o. c.	Wall board, around entire edge ....
1	15½	1010	3" o. c.	Wall board, intermediate nailings ......

penny of the nail should not be greater than the thickness, in eighths of an inch, of the board into which the nail is being driven.

2. In very soft woods, the nails may be one penny larger, or even in some cases, two penny larger.

3. In hard woods, nails should be one penny smaller.

The kind of wood is, of course, a big factor in determining the size of nail to use. The dry weight of the wood is the best basis for the determination of its grain substance or strength. The greater its dry weight, the greater its power to hold nails. However, the splitting tendency of hard wood tends to offset its additional holding power. Smaller nails can be used in hard timber than in soft timber, as shown in Fig. 11. Positive rules governing the size of nails to be used as related to the density of the wood cannot be laid down. Experience is the best guide.

WIRE NAILS—KINDS AND QUANTITIES REQUIRED

The following example of Table 7 will illustrate its usefulness:

Example—What size, type, and quantity of nails are required to lay 1" × 3" flooring for a hall 50' × 100', with joists spaced 16 inches on centers.

Quantities Required (continued)
and Wire Co.)

Trade Names	Pounds per 1000 feet B. M. on center as follows:				
	12″	16″	20″	36″	48″
	Pounds				
American felt roofing	1½ pounds to the square.				
American felt roofing	3 pounds to the square.				
3d shingle	4½ pounds; about 2 nails to each 4 inches.				
4d shingle	7½ pounds; about 2 nails to each 4 inches.				
American felt roofing	12 lbs., ⅝″ heads; 4 nails to shingle.				
Barbed roofing	4½ lbs., ⅝″ heads; 4 nails to shingle.				
2d Barbed Berry, flat head	5 pounds, ⅝″ heads; per 1000 square feet.				
2d casing or flooring	2½ lbs.; ⅝″ heads; per 1000 square feet.				

† Wood shingles vary in width; asphalt are usually 8 inches wide. Regardless of width, 1000 shingles are the equivalent of 1000 pieces 4 inches wide.

Look in the fifth column, headed "Sizes and Kinds of Materials," and find "Flooring, 1 × 3." Follow the line to the right; the size and type specified in the "Trade Names" column is 8d flooring.

Continue on the same line. In the column for 16-inch centers under "Pounds per 1000 feet B.M.," it is found that 32 pounds of nails are required per 1000 feet B.M.

$$B.M. \text{ (board measure) for flooring 1 inch thick} =$$
$$50 \times 100 = 5000$$

$$\text{quantity of nails} = 32 \times \frac{5000}{1000} = 160 \text{ pounds}$$

DRIVING NAILS

One advantage of wire nails is that it is not necessary to hold them in a certain position when driving them to prevent splitting. However, in some instances it is advisable to first drill holes nearly the size of the nail before driving, to guard against splitting. Also, in fine work, where a large number of nails must be driven, such as in boat building, holes should be driven. This step prevents crushing the wood and possible splitting because of the large number of nails driven through each plank. The size of drill for a given size nail should be found by experiment.

The right and wrong ways to drive a nail are shown in Fig. 12. Fig. 13 illustrates the necessity of using a good hammer to drive a nail. The force that drives the nail is due to the inertia of the hammer. This inertia depends on the suddenness with which its motion is brought to rest on striking the nail. With hardened steel, there is practically no give, and all the energy possessed by the hammer is transferred to the nail. With soft and/or inferior metal,

Table 8. Approximate Number

American Steel & Wire Co's. Steel Wire Gauge	Length									
	³⁄₁₆	¼	³⁄₈	½	⅝	¾	⅞	1	1⅛	1¼
⅜	....	....	....	....	....	....	....	29	26	23
⁵⁄₁₆	....	....	....	....	....	....	....	43	38	34
1	....	....	....	....	....	....	....	47	44	40
2	....	....	....	....	....	....	....	60	54	48
3	....	....	....	....	....	....	....	67	60	55
4	....	....	....	....	....	....	....	81	74	66
5	....	....	....	....	....	....	....	90	81	74
6	....	....	....	213	174	149	128	113	101	91
7	....	....	....	250	205	174	148	132	120	110
8	....	....	....	272	238	198	174	153	139	126
9	....	....	....	348	286	238	213	185	170	152
10	....	....	....	469	373	320	277	242	616	196
11	....	....	....	510	117	366	323	285	254	233
12	....	....	....	740	603	511	442	397	351	327
13	....	....	1356	1017	802	688	590	508	458	412
14	....	2293	1664	1290	1037	863	765	667	586	536
15	....	2899	2213	1619	1316	1132	971	869	787	694
16	....	3932	2770	2142	1708	1414	1229	1099	973	872
17	....	5316	3890	2700	2306	1904	1581	1409	1253	1139
18	....	7520	5072	3824	3130	2608	2248	1976	1760	1590
19	....	9920	6860	5075	4132	3508	2816	2556	2284	2096
20	18620	14050	9432	7164	5686	4795	4230	3596	3225	2893
21	23260	17252	12000	8920	7232	6052	5272	4576	4020	3640
22	28528	21508	14676	11776	9276	7672	....	....	....	....
23	35864	27039	18026	13519	10815	9013	....	....	....	....
24	44936	34018	22678	17008	13607	11339	....	....	....	....
25	57357	43243	28828	21622	17297	14414	....	....	....	....

all the energy is not transferred to the nail; therefore, the drive per blow is less than with hardened steel.

Nails for Hardwood Boxes

3/8	in. thickness use 4*d* cement coated nails
7/16 or 1/2	" " " 5*d* " " "
9/16 or 5/8	" " " 6*d* " " "
7/8	" " " 7*d* " " "

Nails for Softwood Boxes

1/4	in. thickness use 4*d* cement coated nails
3/8	" " " 5*d* " " "
7/16 or 1/2	" " " 6*d* " " "
9/16 or 5/8	" " " 7*d* " " "
7/8	" " " 8*d* " " "

1/4 in. thickness use special large 3*d* or regular 4*d* cement coated nails

of Wire Nails per Pound

American Steel & Wire Co's. Steel Wire Gauge	Length										
	1½	1¾	2	2¼	2½	2¾	3	3½	4	4½	5
⅜	20	17	15	15	12	11	11	8.9	7.9	7.1	6.4
5⁄16	29	25	22	20	18	16	15	13	11	10	9.0
1	34	29	26	23	21	20	18	16	14	12	11
2	41	35	31	28	25	23	21	18	16	14	13
3	47	41	36	32	29	27	25	21	18	16	15
4	55	48	41	37	34	31	29	25	22	20	18
5	61	52	45	41	38	35	32	28	24	22	21
6	76	65	58	52	47	43	39	34	29	26	24
7	92	78	70	61	55	51	47	40	35	31	28
8	106	93	82	74	66	61	56	48	42	38	34
9	128	112	99	87	79	71	67	58	50	45	41
10	165	142	124	111	100	91	84	71	62	55	49
11	200	171	149	136	122	111	103	87	77	69	61
12	268	229	204	182	161	149	137	118	103	95	87
13	348	297	260	232	209	190	175	153	138	123	110
14	459	398	350	312	278	256	233	201	176	157	140
15	578	501	437	390	351	317	290	256	220	196	177
16	739	635	553	496	452	410	370	318	277	248	226
17	956	831	746	666	590	532	486	418	360	322	295
18	1338	1150	996	890	820	740	680	585	507	448	412
19	1772	1590	1390	1205	1060	970	895	800	....	....	....
20	2412	2070	1810	1620	1450	1315	1215	1035	....	....	....
21	3040	2665	2310	2020	1830	....	....	....	....	....	....
22	....	....	....	....	....	....	....	....	....	....	....
23	....	....	....	....	....	....	....	....	....	....	....
24	....	....	....	....	....	....	....	....	....	....	....
25	....	....	....	....	....	....	....	....	....	....	....

These approximate numbers are an average only, and the figures given may be varied either way, by changes in the dimensions of the heads or points. Brads and on-head nails will run more to the pound than table shows, and large or thick-headed nails will run less.

AXIS OF HANDLE PARALLEL WITH PLANE OF NAIL HEAD

USUALLY 10° TO 45°

HAMMER STRIKING SQUARE ON HEAD - FULL CONTACT

RIGHT WAY

WRONG WAY

Fig. 12. Right and wrong ways to drive a nail. Hit the nail squarely on the head. The handle should be horizontal when the hammer head hits a vertical nail.

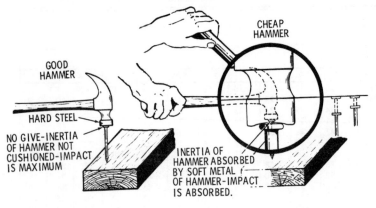

CHEAP HAMMER

GOOD HAMMER

HARD STEEL

NO GIVE-INERTIA OF HAMMER NOT CUSHIONED-IMPACT IS MAXIMUM

INERTIA OF HAMMER ABSORBED BY SOFT METAL OF HAMMER-IMPACT IS ABSORBED.

Fig. 13. Why a cheap hammer should not be used.

SUMMARY

Nails are the carpenters' most useful fastener, and a great variety of types and sizes are available to meet the demands of the industry. On any kind of construction work, an important consideration is the type and size of nails to use.

The first factor to consider is finish. Should the nail be smooth, barbed, or cement coated? The holding power of cement-coated nails, is considerably greater than that of the same size smooth nail. The second factor to consider is the diameter of the nail. Long, thin nails will break at the joints of the lumber. Short, thick nails will work loose quickly. The kind of wood is, of course, a big factor in determining the size of nail to use.

REVIEW QUESTIONS

1. What is nail holding power?
2. What type of nail is considered best in holding power?
3. Explain the "penny" nail system.
4. What should be considered when selecting a nail for a particular job?
5. What type of common nail has the best holding power?

CHAPTER 3

Screws

Wood screws are not often used in structural carpentry because their advantage over nails is only in their greater withdrawal resistance, and structural loadings that place the fastenings in withdrawal are always avoided when possible. Their lateral or shearing resistance is not appreciably greater than that of driven nails of the same diameters. In some types of cabinet work, screws are used to allow ready dismounting, or disassembly, and their appearance is good. They are used in installing all kinds of builder's hardware, because of their great withdrawal resistance and because they are more or less readily removed in case repairs or alterations are necessary.

By definition, a wood screw is a fastening implement consisting of a shank with a rather coarse, sharp, right-hand thread, a sharp gimlet point that enters the wood readily, and a head that may be any one of various contours, and which is provided with a slot, or crossed slots, to receive the tip of a driving tool.

Screws of many types are made for specialized purposes, but stock wood screws are usually obtainable in either steel or brass, and, more rarely, of high strength bronze. Three types of heads are standard: the flat countersunk head, with the included angle of the sloping sides standardized at 82°; the round head, whose height is also standardized, but whose contour seems to vary slightly among the products of different manufacturers; and the oval head, which combines the contours of the flat head and the round head. All of these screws are available with the Phillips slot, or crossed slots, instead of the usual single straight slot.

The Phillips slot allows a much greater driving force to be exerted without damaging the head, and it is more sightly than the usual straight-slotted head. By far the greater part of all wood screws used, probably 75% or more, are of the flat-head type.

MATERIAL

For ordinary purposes, steel screws, with or without protective coatings, are commonly used. In boat building or other such work where corrosion will probably be a problem if screws are used, the screws should be of the same metal or at least the same *type* of metal as the parts they contact. While it is possible and indeed probable that a single brass screw that is driven through an aluminum plate, if it is kept dry, will show no signs of corrosion, many brass screws driven through the aluminum plate in the presence of water or dampness will almost certainly show signs, perhaps serious, of galvanic corrosion.

DIMENSIONS OF SCREWS

When ordering screws, it is important to know what constitutes the length of a screw. The overall length of a 2-inch flat-head screw is not the same as that of a 2-inch round-head screw. To avoid confusion and mistakes, the lengths for various types of screws are shown in Fig. 1 and should be carefully noted.

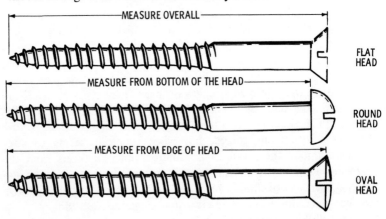

Fig. 1. Various wood screws and how their length is measured.

It should also be noted that, unlike the ordinary wire gauges, the 0 in the screw gauge, shown in Fig. 2, indicates the diameter of the smallest screw, and the diameter of the screws *increase* with the number of the gauge.

SHAPE OF THE HEAD

The buyer will find a multiplicity of head shapes to select from; the variety of heads regularly carried in a well-stocked hardware store is usually great enough to meet every possible requirement. However, in order to avoid possible disappointment where the supply base (small dealers) is remote from large centers, it is better to select from these three forms of heads, which may be regarded as standard:

1. Flat
2. Round
3. Oval

All of these heads are available in either the straight-slotted or Phillips type.

The other forms may be regarded as special or semispecial, that is, carried by large dealers only or obtainable only on special order.

Flat heads are necessary in some cases, such as on door hinges, where any projection would interfere with the proper working of the hinge; flat-head screws are also employed on finish work where flush surfaces are desirable. The round and oval heads are normally ornamental when exposed.

The diameter of the head in relation to the gauge number of the screw is shown in Table 1. Some of the many special screw heads available are shown in Fig. 3.

HOW TO DRIVE A WOOD SCREW

Consult Table 2 to determine the size of drill to use in drilling the shank-clearance hole. This hole (Fig. 4) should be slightly smaller than the shank diameter of the screw and about ¾ the shank length for soft and medium-hard woods. For extremely hard woods, the length of the hole should equal the shank length.

53

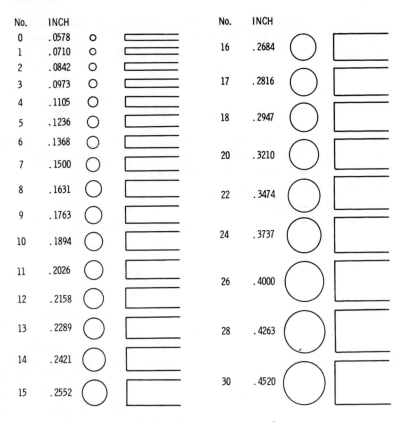

No.	INCH		
0	.0578		
1	.0710		
2	.0842		
3	.0973		
4	.1105		
5	.1236		
6	.1368		
7	.1500		
8	.1631		
9	.1763		
10	.1894		
11	.2026		
12	.2158		
13	.2289		
14	.2421		
15	.2552		

No.	INCH		
16	.2684		
17	.2816		
18	.2947		
20	.3210		
22	.3474		
24	.3737		
26	.4000		
28	.4263		
30	.4520		

Fig. 2. Wood screw gauge numbers.

Again consult Table 2 to determine the size of drill to use in drilling the pilot hole. This hole should be equal in diameter to the root diameter of the screw thread and about ¾ the thread length for soft and medium-hard woods. For extremely hard woods, the pilot hole depth should equal the thread length.

If the screw being inserted is the flat-head type, the hole should be countersunk. A typical countersink is shown in Fig. 5.

The foregoing process involves three separate steps. All of these can be performed at once by using a device of the type shown in Fig. 6. This tool will drill the pilot hole, the shank-clearance hole, and the countersink all in one operation.

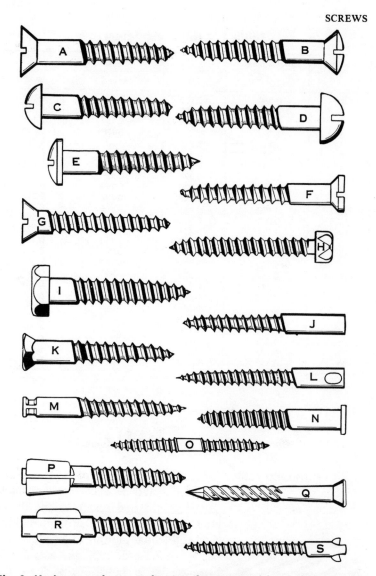

Fig. 3. Various wood screws showing the variety of head shapes available. A, flat head; B, oval head; C, round head; D, piano head; E, oval fillister head; F, countersunk fillister head; G, felloe; H, close head; I, hexagon head; J, headless; K, square bung head; L, grooved; M, pinched head; N, round bung head; O, dowel; P, winged; Q, drive; R, winged; S, winged head. Heads A through G may be obtained with Phillips-type head.

Table 1. Head Diameters

Screw Gauge	Screw Diameter	Head Diameter		
		Flat	Round	Oval
0	0.060	0.112	0.106	0.112
1	0.073	0.138	0.130	0.138
2	0.086	0.164	0.154	0.164
3	0.099	0.190	0.178	0.190
4	0.112	0.216	0.202	0.216
5	0.125	0.242	0.228	0.242
6	0.138	0.268	0.250	0.268
7	0.151	0.294	0.274	0.294
8	0.164	0.320	0.298	0.320
9	0.177	0.346	0.322	0.346
10	0.190	0.371	0.346	0.371
11	0.203	0.398	0.370	0.398
12	0.216	0.424	0.395	0.424
13	0.229	0.450	0.414	0.450
14	0.242	0.476	0.443	0.476
15	0.255	0.502	0.467	0.502
16	0.268	0.528	0.491	0.528
17	0.282	0.554	0.515	0.554
18	0.394	0.580	0.524	0.580
20	0.321	0.636	0.569	0.636
22	0.347	0.689	0.611	0.689
24	0.374	0.742	0.652	0.742
26	0.400	0.795	0.694	0.795
28	0.426	0.847	0.735	0.847
30	0.453	0.900	0.777	0.900

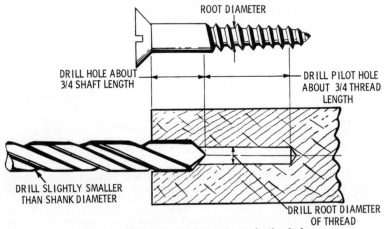

Fig. 4. Drilling shank-clearance and pilot holes.

Fig. 5. A typical countersink.

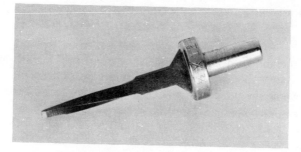

Fig. 6. A tool for drilling pilot hole, shank-clearance hole, and counter-sink in one operation.

Fig. 7. Ordinary lag screw.

These tools are made in many sizes, one for each screw size, and they are available in complete sets or separately. The screw size is marked on the tool.

STRENGTH OF WOOD SCREWS

Table 2 gives the safe resistance, or safe load (against pulling out), in pounds per linear inch of wood screws when inserted across the grain. For screws inserted with the grain, use 60% of these values.

57

Table 2. Safe Loads for Wood Screws

Kind of Wood	Gauge Number							
	4	8	12	16	20	24	28	30
White oak	80	100	130	150	170	180	190	200
Yellow pine	70	90	120	140	150	160	180	190
White pine	50	70	90	100	120	140	150	160

The lateral load at right angles to the screw is much greater than that of nails. For conservative designing, assume a safe resistance of a No. 20 gauge screw at double that given for nails of the same length, when the full length of the screw thread penetrates the supporting piece of the two connected pieces.

LAG SCREWS

By definition, a lag screw, as shown in Fig. 7, is a heavy-duty wood screw that is provided with a square or hexagon head so that it may be turned by a wrench. These are large, heavy screws that are used where great strength is required, such as for heavy timber work, etc. Table 3 gives the dimensions of ordinary lag screws.

Table 3. Lag Screws

Length	3	3½	4	4½	5	5½	6	6½	7	7½	8	9	10	11	12
Dia.	5/16 to 7/8	5/16 to 1	5/16 to 1	5/16 to 1	5/16 to 1	5/16 to 1	5/16 to 1	7/16 to 1	7/16 to 1	7/16 to 1	7/16 to 1	7/16 to 1	1/2 to 1	1/2 to 1	1/2 to 1

How to Put in Lag Screws

First, bore a hole slightly larger than the diameter of the shank to a depth that is equal to the length that the shank will penetrate (see Fig. 8). Then bore a second hole at the bottom of the first hole equal to the root diameter of the threaded shank and to a depth of approximately one-half the length of the threaded portion. The exact size of this hole and its depth will, of course, depend on the kind of wood; the harder the wood, the larger the hole.

The resistance of a lag screw to turning is enormous when the hole is a little small, but this can be considerably decreased by smearing the threaded portion of the screw with soap or beeswax.

58

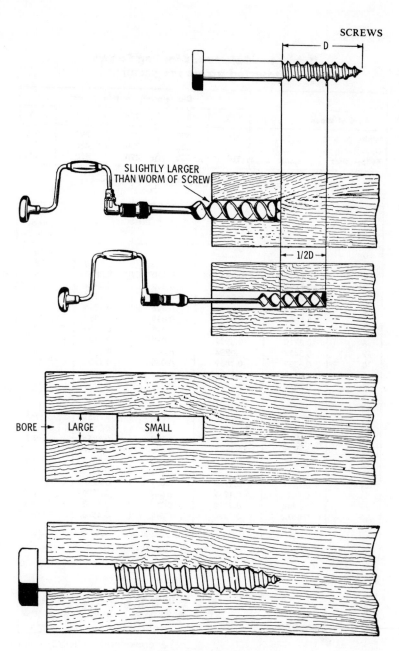

SCREWS

D

SLIGHTLY LARGER
THAN WORM OF SCREW

1/2D

BORE — LARGE   SMALL

Fig. 8. Drilling holes for lag screws.

Table 4.   Safe Loads for Lag Screws
(Inserted across the grain)

Kind of Wood	Diameter of Screw in Inches				
	1/2	5/8	3/4	7/8	1
White pine	590	620	730	790	900
Douglas fir	310	330	390	450	570
Yellow pine	310	330	390	450	570

Table 5.   Standard Wood Screw Proportions

Screw Numbers	A	B	C	D	Number of Threads per Inch
0	....	....	....	0.0578	30
1	....	....	....	0.0710	28
2	0.1631	0.0454	0.030	0.0841	26
3	0.1894	0.0530	0.032	0.0973	24
4	0.2158	0.0605	0.034	0.1105	22
5	0.2421	0.0681	0.036	0.1236	20
6	0.2684	0.0757	0.039	0.1368	18
7	0.2947	0.0832	0.041	0.1500	17
8	0.3210	0.0809	0.043	0.1631	15
9	0.3474	0.0984	0.045	0.1763	14
10	0.3737	0.1059	0.048	0.1894	13
11	0.4000	0.1134	0.050	0.2026	12.5
12	0.4263	0.1210	0.052	0.2158	12
13	0.4427	0.1286	0.055	0.2289	11
14	0.4790	0.1362	0.057	0.2421	10
15	0.5053	0.1437	0.059	0.2552	9.5
16	0.5316	0.1513	0.061	0.2684	0
17	0.5579	0.1589	0.064	0.2815	8.5
18	0.5842	0.1665	0.066	0.2947	8
20	0.6368	0.1816	0.070	0.3210	7.5
22	0.6895	0.1967	0.075	0.3474	7.5
24	0.7421	0.2118	0.079	0.3737	7
26	0.7421	0.1967	0.084	0.4000	6.5
28	0.7948	0.2118	0.088	0.4263	6.5
30	0.8474	0.2270	0.093	0.4546	6
....	....	....	....	....	....

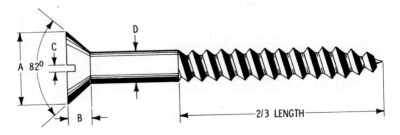

Strength of Lag Screws

Table 4 gives the safe resistance, or load to pulling out, in pounds per linear inch of thread for lag screws when inserted across the grain.

SUMMARY

Wood screws are often used in carpentry because of their advantage over nails in strength. They are used in installing various types of building hardware, because of their great withdrawal resistance and because they are more or less readily removed in case of repairs or alterations.

There are generally three standard types of screw heads—the flat countersunk head, the round head, and the oval head, all of which can be obtained in crossed slot, single straight slot, or Phillips slot. For ordinary purposes, steel screws are commonly used, and in wood-screw applications, probably 75 percent or more are flathead screws.

Lag screws are heavy-duty wood screws that are provided with a square or hexagonal head so that they may be installed with a wrench. These are large, heavy screws that are used where great strength is needed, such as for heavy timber and beam installations. Holes are generally bored into the wood since the diameter of lag screws is large.

REVIEW QUESTIONS

1. Name the three basic head shapes of wood screws.
2. What are the three screw slots used on wood screws?
3. What type of wood screw is used where great strength is required?
4. What type of head is used on lag screws? Why?
5. What is meant by the root diameter of a screw?

CHAPTER 4

Bolts

By definition, a bolt is a pin or rod that is used for holding anything in its place, and often having a permanent head on one end. A bolt is generally regarded as a rod having a head at one end and threaded at the other to receive a nut; the nut is usually considered as forming a part of the bolt, and prices quoted on bolts by dealers normally include the nuts.

KINDS OF BOLTS

There is a multiplicity of bolt forms to meet various requirements.

The *common machine bolt* has a square head at one end and a short length of thread at the other end. Two of these are shown in Fig. 1.

When a loop, or "eye," is provided instead of a head, it is called an *eye bolt*.

A *countersunk bolt* has a beveled head, which fits into a countersunk hole.

A *key-head bolt* has a head so shaped that, when inserted into a suitable groove or slot provided for it, it will not turn when the nut is screwed onto the other end.

Another method of preventing turning consists of forming a short portion of the bolt body square at the head end, the head itself being spherical in shape; such a type is known as a *carriage bolt*.

A headless bolt threaded for a certain distance at both ends is called a *stud bolt*.

In addition to the types just mentioned, there are numerous others, such as *milled coupling, railroad track, stove, expansion,* etc. Several different types of bolts are illustrated in Fig. 3.

BOLTS

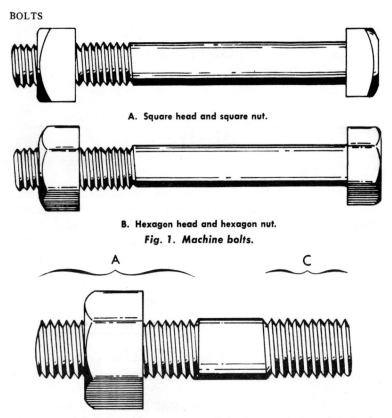

A. Square head and square nut.

B. Hexagon head and hexagon nut.

Fig. 1. Machine bolts.

A   C

*Fig. 2. Stud bolt with hexagon nut. A is the nut end and C is the
attachment end.*

MANUFACTURE OF BOLTS

The bolt-and-nut industry in America was started on a small
scale in Marion, Connecticut, in 1818. In that year, Micah Rugg,
a country blacksmith, made bolts by the forging process. The first
machine used for this purpose was a device known as a heading
block, which was operated by a foot treadle and a connecting
lever. The connecting lever held the blank while it was being
driven down into the impression in the heading block by a hammer. The square iron from which the bolt was made was first
rounded, so that it could be admitted into the block.

At first, Rugg only made bolts to order, and charged at the rate
of 16 cents apiece. This industry developed quite slowly until

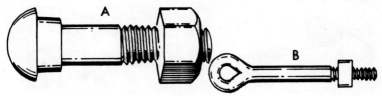

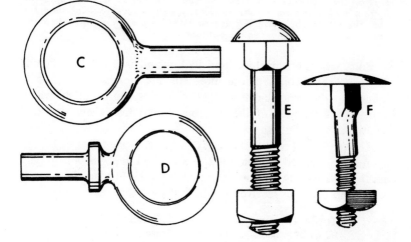

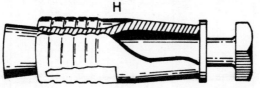

Fig. 3. Various bolts. In the figure, A is a railroad track bolt; B, a welded eye bolt; C, a plain forged eye bolt; D, a shouldered eye bolt; E, a carriage bolt; F, a step bolt; G, a stove bolt; H, an expansion bolt.

1839, when Rugg went into partnership with Martin Barnes. Together they built the first exclusive bolt-and-nut factory in the United States at Marion, Connecticut.

65

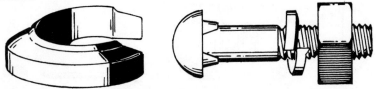

Fig. 4. The National lock washer. When the nut is screwed onto the bolt, it strikes the rib on the washer, which is much harder than the nut. The rib on the washer is forced into the nut, thus preventing the nut from loosening.

Bolts were first manufactured in England in 1838 by Thomas Oliver of Darlston, Staffordshire. His machine was built on a somewhat different plan from that of Rugg's, but no doubt was a further development of the first machine. Oliver's machine was known as the "English Oliver."

The construction of the early machines was carefully kept secret. It is related than in 1842, a Mr. Clark had his bolt-forging machine located in a room separated from the furnaces by a thick wall. The machine received the heated bars through a small hole

Table 1. Properties of U.S. Standard Bolts
(U.S. Standard or National Coarse Threads)

Diameter	Number of Threads per inch (National Coarse Thread)	Head	Head	Head
1/4	20	3/8	13/32	1/2
5/16	18	1/2	35/64	43/64
3/8	16	9/16	5/8	3/4
7/16	14	5/8	11/16	53/64
1/2	13	3/4	53/64	1
9/16	12	7/8	31/32	1 5/32
5/8	11	15/16	1 1/32	1 1/4
3/4	10	1 1/8	1 15/64	1 1/2
7/8	9	1 5/16	1 29/64	1 47/64
1	8	1 1/2	1 21/32	1 63/64
1 1/8	7	1 11/16	1 55/64	2 15/64
1 1/4	7	1 7/8	2 1/16	2 31/64
1 3/8	6	2 1/16	2 17/64	2 47/64
1 1/2	6	2 1/4	2 31/64	2 63/64
1 5/8	5 1/2	2 7/16	2 11/16	3 15/64
1 3/4	5	2 5/8	2 57/64	3 31/64
1 7/8	5	2 13/16	3 3/32	3 47/64
2	4 1/2	3	3 5/16	3 63/64

cut in the wall; the forge man was not even permitted to enter the room.

A modern bolt-and-rivet machine consists of two gripping dies, one movable and the other stationary, and a ram which carries the heading tool. The heated bar is placed in the impression in the stationary gripping die, and against the gauge stop. The machine is then operated by pressing down a foot treadle. On this type of machine, the bar is generally cut to the desired length before heading, especially when it is long enough to be conveniently gripped with the tongs, but it can be headed first and afterwards cut off to the desired length. It is also possible in some makes of machines to insert a cutting tool that can cut off the blank before heading, when the work is not greater in length than the capacity of the machine.

PROPORTIONS AND STRENGTH OF BOLTS

Ordinary bolts are manufactured in certain "stock sizes." Table 1 gives these sizes for bolts from ¼" up to 1¼", with the length of thread.

For many years, the coarse-thread bolt was the only type available. In recent years, bolts with a much finer thread, called the National Fine thread, have become easily available. These have hex heads and hex nuts. They are much better finished than the stock coarse-thread bolts and consequently, are more expensive. Cheap rolled-thread bolts, with the threaded portions slightly upset, should not be used by the carpenter. When they are driven into a hole, either the hole is too large for the body of the bolt or the threaded portion reams it out too large for a snug fit. Good bolts

Table 2. National Fine Threads

Diameter	Threads per inch
1/4	28
5/16	24
3/8	24
7/16	20
1/2	20
9/16	18
5/8	18
3/4	16
7/8	14
1	14

have cut threads that have a maximum diameter no larger than the body of the bolt.

When a bolt is to be selected for a specific application, Table 3 should be consulted.

Example—How much of a load may be applied to a 1-inch bolt for a tensile strength of 10,000 pounds per square inch?

Referring to Table 3, we find on the line of a 1-inch bolt a value of 5510 pounds corresponding to a stress on the bolt of 10,-000 pounds per square inch.

Table 3.   Proportions and Strength of U.S. Standard Bolts

Bolt Diameter	Area at Bottom of Threads	Tensile Strength		
		10,000 lbs/in²	12,500 lbs/in²	17,500 lbs/in²
1/4	0.027	270	340	470
5/16	0.045	450	570	790
3/8	0.068	680	850	1190
7/16	0.093	930	1170	1630
1/2	0.126	1260	1570	2200
9/16	0.162	1620	2030	2840
5/8	0.202	2020	2520	3530
3/4	0.302	3020	3770	5290
7/8	0.419	4190	5240	7340
1	0.551	5510	6890	9640
1 1/8	0.693	6930	8660	12,130
1 1/4	0.890	8890	11,120	15,570
1 3/8	1.054	10,540	13,180	18,450
1 1/2	1.294	12,940	16,170	22,640
1 5/8	1.515	15,150	18,940	26,510
1 3/4	1.745	17,450	21,800	30,520
1 7/8	2.049	20,490	25,610	35,860
2	2.300	23,000	28,750	40,250

Example—What size bolt is required to support a load of 4000 pounds for a stress of 10,000 pounds per square inch?

$$area\ at\ root\ of\ thread = given\ load \div 10,000$$
$$= 4000 \div 10,000 = 0.400\ sq.\ in.$$

Referring to Table 3, in the column headed "Area at Bottom of Thread," we find 0.419 square inch to be the nearest area; this corresponds to a ⅞-inch bolt.

Of course, for the several given values of pounds stress per square inch, the result could be found directly from the table, but

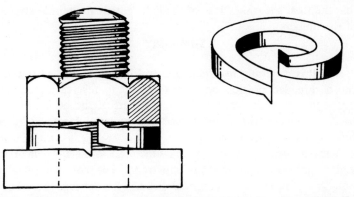

Fig. 5. Positive lock washer. The barbs force themselves deeply into the nut and the backing.

the calculation above illustrates the method that would be employed for other stresses per square inch not given in the table.

Example—A butt joint with fish plates is fastened by six bolts through each timber. What size bolts should be used, allowing a shearing stress of 5000 pounds per square inch in the bolts, when the joint is subjected to a tensile load of 20,000 pounds?

$$load\ carried\ per\ bolt = 20,000 \div number\ of\ bolts$$
$$= 20,000 \div 6 = 3333\ lbs.$$

Each bolt is in double shear, hence:

$$equivalent\ single\ shear\ load = \tfrac{1}{2}\ of\ 3333 = 1667\ lbs.$$

and

$$area\ per\ bolt = \frac{1667}{5000} = 0.333\ sq.\ in.$$

Referring to Table 3, the nearest area is 0.302, which corresponds to a ¾-inch bolt. In the case of a dead, or "quiescent," load, ¾-inch bolts would be ample; however, for a live load, take the next larger size, or ⅞-inch bolts.

The example does not give the size of the timbers, but the assumption is they are large enough to safely carry the load. In practice, all parts should be calculated as described in the chapter on the strength of timbers. The ideal joint is one so proportioned that the total shearing stress of the bolts equals the tensile strength of the timbers.

SUMMARY

A bolt is generally regarded as a rod having a head at one end and a threaded portion on the other end to receive a nut. The nut is usually considered as forming a part of the bolt. Bolts are used to connect two or more pieces of material together thereby forming a strong connection.

Various forms of bolts are manufactured to meet the demands and requirements of the building trade. The common machine bolt has a square or hexagonal head. The carriage bolt has a round head; the stove bolt has a round or countersunk head with a single slot. Lock washers are used to prevent the nut from loosening.

REVIEW QUESTIONS

1. What type of head is generally found on a machine bolt?
2. What is meant by "threads per inch"?
3. Explain the purpose of lock washers.
4. What is an expansion bolt?
5. What is tensile load?

CHAPTER 5

The Workbench

In order to properly perform many of the numerous operations in carpentry, a suitable workbench is essential. Broadly speaking, the bench may be regarded as a tool, especially when considered

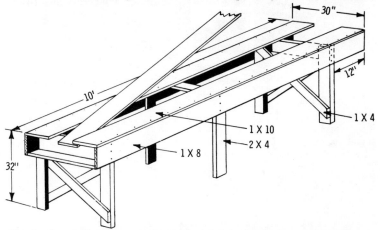

Fig. 1. A quickly constructed workbench for temporary use. Lumber required: five 2" x 4" x 8'; one 1" x 4" x 16'; two 1" x 8" x 10'; and three 1" x 10" x 10'. The center legs add rigidity, but on such a short bench, they are not absolutely necessary.

with its various attachments for holding and clamping the material during the numerous operations known as benchwork.

There is hardly a shop for any purpose that does not require a workbench, especially the carpenter's shop. Many of these shops have only a makeshift bench made of $2'' \times 4''$ and 1-inch boards, crudely put together, with no part of it strong enough and no place to attach a vise; such benches temporarily knocked together on the job are not worthy to be called workbenches.

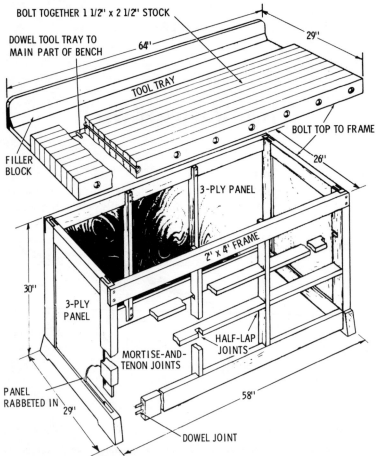

BOLT TOGETHER 1 1/2" x 2 1/2" STOCK

DOWEL TOOL TRAY TO
MAIN PART OF BENCH

64"

29"

TOOL TRAY

BOLT TOP TO FRAME

26"

FILLER
BLOCK

3-PLY PANEL

2' x 4' FRAME

30"

3-PLY
PANEL

HALF-LAP
JOINTS

MORTISE-AND-
TENON JOINTS

PANEL
RABBETED IN   29"

58"

DOWEL JOINT

Fig. 2. Construction details of a workbench designed for appearance as well as convenience. The 2" x 4" rails, corner posts, and base members provide a sturdy bench that will give many years of satisfactory service.

Substantial benches are manufactured and sold for all purposes, but a carpenter can construct for himself the type of bench best suited to his requirements.

Fig. 1 shows a quickly constructed bench such as would be used at a construction site. A bench such as that shown in Fig. 2 should be constructed with more care for permanent installation in a shop.

The height of the bench should be regulated by the character of the work to be done—high for light work and low for heavy work. The height of the person who is to use the bench should also be considered. In general, carpenter's benches are made 33 inches high, while those for cabinetmakers and patternmakers are from 2 to 4 inches higher.

WORKBENCH ATTACHMENTS

Numerous devices are used with workbenches to facilitate the operations to be performed. These devices or attachments are:

1. Vises.
2. Support pegs.
3. Bench stop.
4. Bench hook.

Vises

The general construction of a bench vise is shown in Fig. 3. There are numerous types in use. They may be made all of wood, all of iron, or of wood with an iron screw. Usually there is a large, or main, vise of all wood or combined wood and iron construction at the left end of the table (as shown in Fig. 3) and frequently a smaller or supplemental iron vise at the other end for small work.

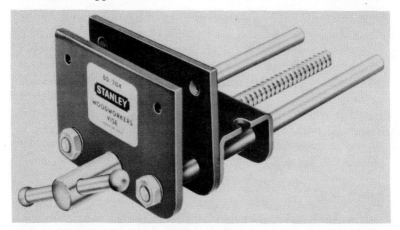

Fig. 3. A bench vise with guides to maintain the jaws parallel. The inner jaw is fastened to the bench and supports a fixed nut in which the screw rotates. The screw moves the outer jaw which has attached to it two rods that slide through guide sleeves to keep the jaws parallel at all times.

For patternmaking work, a quick-closing vise is frequently used. It is advisable to face any iron vise with a wood or leather covering to prevent marking or denting the lumber, especially when soft woods are used.

Support Pegs

The function of the main bench vise is to prevent the wood from moving while being worked, as with a plane or chisel. In these oper-

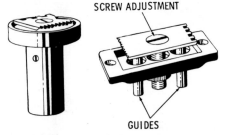

SCREW ADJUSTMENT

GUIDES

Fig. 4. Round and rectangular forms of a bench stop. These adjust by a center screw from flush to as high as required for the work. The round bench stop is fitted by boring a hole the diameter of the stop with an expansion bit and a deeper center with the proper size of bit. The rectangular bench stop is shallow and should be mortised in flush with the bench top.

ations, the wood receives a pressure that tends to rotate it in the plane of the vise jaws, the latter acting as a pivot.

In the case of a long board, this turning force, or torque, would become extremely great when a downward pressure is applied at the far end of the board, thereby requiring the vise to be screwed up rather firmly to prevent turning. To avoid this, the bench is provided with supporting pegs, which carry the weight of the board and prevent it from turning when a downward pressure is applied in tooling. A vertical row of holes for the pegs should be provided at the middle and at the right end of the bench.

Bench Stop

This device is intended to prevent any longitudinal movement of the work while it is being tooled; that is, it prevents endwise movement of a board while the board is being planed. As usually constructed for this purpose, a bench stop (Fig. 4) consists of a metal casing that is designed to set in flush with the bench and has a

74

horizontal toothed plate that works in vertical guides. A screw adjustment is provided so that the plate may be set flush with the top of the table (when not in use) or a little above so as to engage the end of the work and thus prevent endwise movement.

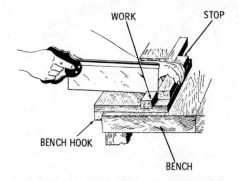

Fig. 5. Typical bench hook and method of use when sawing to size with back saw. The bench hook is used for a variety of operations, such as odd sawing and chiseling, and also serves to prevent the workbench from being marred by such operations.

Bench Hook

This is virtually a moveable stop that can be used at right angles to the front of the bench. It serves many purposes for holding and putting work together. When it is desired to saw off a piece of stock, the bench hook is placed on the bench (Fig. 5); one shoulder is set against the edge of the bench, and the upper shoulder serves as a stop for the work while sawing.

Tool Panels

These, as the name implies, provide assigned positions for each tool and will induce the craftsman to keep his shop orderly by putting the tools back where they belong after each use.

A place for everything and everything in its place should be the aim of every home workshop operator. If it is assumed that a tool collection is available for mounting, the best system is to plan and fit the various units properly before the panel is put in place. The simplest tool mounting is by means of a pair of nails, although a neater and more satisfactory method is to use wood or metal dowels. The usual practice is to have one or two racks with holes or slots, while the design of the rest of the panel

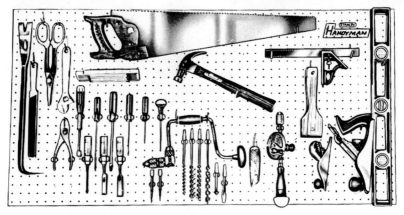

Fig. 6. A tool panel showing arrangement of tools.

depends on the number of tools available and their size and frequency of use. Perforated hardboard with specially designed tool hangers, as illustrated in Fig. 6, can be purchased.

Prior to mounting the panel it should be given a coat or two of shellac or paint. This will give the panel a neat and clean appearance. Tool panels of this type are usually made of hardboard and are strong enough to support the combined weight of the tools mounted on them.

SUMMARY

There is hardly a worshop that does not require a workbench of some kind, especially the woodworking shop. Substantial benches are manufactured and sold for all purposes, but a carpenter or the do-it-yourself person can construct the type of bench best suited to his requirements.

A bench should be constructed with care as a permanent installation in the shop. The height and width of the bench should be regulated by the character of the work to be done—high enough for light work, and low enough for heavy work. The height of the person who is going to use the bench should also be considered.

The general construction of a bench should be heavy enough, or at least anchored to the floor, in order to withstand hard rough work without moving. Various bench attachments or devices, such as vises, bench stops, and bench hooks, are used with workbenches to facilitate the operations to be performed.

REVIEW QUESTIONS

1. What size lumber should be used when making a good sturdy workbench?
2. What are workbench attachments? Name a few.
3. Why should a workbench be heavy or at least anchored to the floor?
4. Why should you always have a tool panel?

CHAPTER 6

Carpenters' Tools

A carpenter should possess a full set of tools; in the selection of these, it is important to buy only the best, regardless of cost. Select carefully from standard makes, examining them carefully to be sure there are no visible defects. The temper of steel may be discovered only by use, and any defect in the best grades of tools is normally made good upon complaint to the dealer; therefore, buy only the best.

When classifying the multiplicity of carpenters' tools, they should be grouped according to some particular logical system that will divide them into well-defined groups. Therefore, with respect to use, tools may be classified as:

1. Guiding and testing tools.
   Straightedge
   Square
      try square
      miter square
      combined try and miter square
      framing or so-called "steel" square
      combination square
   Sliding "T" bevel
   Miter box
   Level
   Plumb bob
   Plumb rule

2. Marking tools.
   Chalk line
   Carpenter's pencil
   Ordinary pencil

    Scratch awl
    Scriber
    Compass and dividers

3. Measuring tools.
    Carpenter's two-foot rule
    Various folding rules
    Rules with attachments
    Lumber scales
    Marking gauges

4. Holding tools.
    Horses or trestles
    Clamps
    Vises

5. Toothed cutting tools.
    Saws
        hand
        circular
        band
    Files and rasps
    Sandpaper

6. Sharp-edged cutting tools.
    Chisels
        paring
        firmer
        framing
        slick
        corner
        gouge
        tang and socket
        butt pocket and mill
    Drawknife

7. Rough facing tools.
    Hatchet
    Axe

8. Smooth facing tools.
    Spokeshave
    Planes

jack
fore
jointer
smoothing
block

9. Boring tools.
Awl
Gimlets
Augers
Drills
Hollow augers
Countersinks
Reamers

10. Fastening tools.
Hammers
Screwdrivers
Wrenches

11. Sharpening tools.
Abrasives
Grinding wheels (electric powered)
Oilstones
natural
artificial

SUMMARY

Tools in any trade are a necessity, and it is important to buy only the best, regardless of cost. A careful selection from standard brands, examining them to be sure there are no visible defects, is always a good practice.

The type and number of hand tools required depend on individual preferences and the work to be done, although a good basic list will comprise about 30 essential tools. While the list of hand tools may be increased to a very large number, most ordinary woodworking projects can be performed with a strictly limited number of tools.

REVIEW QUESTIONS

1. What tools are considered measuring tools?
2. Under what category are chalk lines and scratch awls?
3. Name the various types of wood planes.
4. What are tooth cutting tools?
5. What are holding tools?

CHAPTER 7

Guiding and Testing Tools

In good carpentry, much depends on accuracy in measurement and in fitting parts together at the required angle. In order to insure this accuracy, various tools of guidance and direction are used so that joints, etc., can be made with precision.

STRAIGHTEDGE

This tool is used to guide the pencil or scriber when marking a straight line and when testing a faced surface, such as the edge of a board, to determine if it is straight. Anything having an edge known to be straight, such as the edge of a steel square, may be used; however, a regular straightedge is preferable.

The straightedge may be made either of wood or steel, and its length may be from a few inches to several feet. For ordinary work, a carpenter can make a sufficiently accurate straightedge from a strip of good straight-grained wood, as shown in Fig. 1, but for accurate work, a steel straightedge, such as the three shown in Fig. 2, should be used. Wood is objectionable for precision work because of its tendency to warp or spring out of shape.

Fig. 3 shows the correct and incorrect methods of holding a straightedge as a guiding tool, and Fig. 4 shows how and how not to hold the pencil when marking stock.

SQUARE

This tool is a 90°, or right-angle, standard and is used for marking or testing work. There are several common types of squares, as shown in Fig. 5; they are:

    1. Try square.

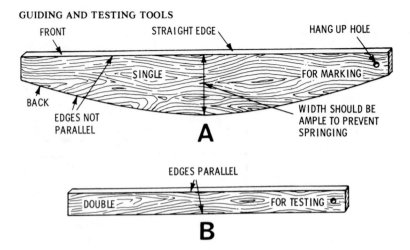

FRONT   STRAIGHT EDGE   HANG UP HOLE

SINGLE   FOR MARKING

BACK

EDGES NOT PARALLEL

WIDTH SHOULD BE AMPLE TO PREVENT SPRINGING

A

EDGES PARALLEL

DOUBLE   FOR TESTING

B

Fig. 1. Wooden straightedges. When well made, they are sufficiently accurate for ordinary use; A, single straightedge; B, double straightedge.

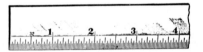

Fig. 2. Typical steel straightedges. These tools are used where straight lines are to be scribed or where surfaces must be tested for flatness. Depending on their use, straightedges are made in lengths of from 12 to 72 inches, widths of 1 3/8 to 2 1/2 inches, and thicknesses of 1/16 to 1/2 inch.

2. Miter square.
3. Combined-try-and-miter square.
4. Framing or so-called "steel" square.
5. Combination square.

Try Square

In England, this is called the trying square, but here it is simply the try square. It is so called probably because of its

Fig. 3. *The incorrect and correct methods of using the straightedge as a guiding tool. To properly secure the straightedge, the hand should press firmly on the tool at its center, with the thumb and other fingers stretched wide apart.*

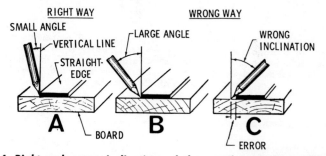

Fig. 4. *Right and wrong inclinations of the pencil in marking with the straightedge. The pencil should not be inclined from the vertical more than is necessary to bring the pencil lead in contact with the guiding surface of the straightedge (A). When the pencil is inclined more, and pressed firmly, considerable pressure is brought against the straightedge, tending to push it out of position (B). If the inclination is in the opposite direction, the lead recedes from the guiding surface, thus introducing an error which is magnified when a wooden straightedge is used because of the greater thickness of the straightedge (C).*

frequent use as a testing tool when squaring up mill-planed stock. The ordinary try square used by carpenters consists of a steel blade set at right angles to the inside face of the stock in which it is held. The stock is made of some type of hardwood and is always faced with brass in order to preserve the wood from injury.

The usual sizes of try squares have blades ranging from 3 to 15 inches long. The stock is approximately ½ inch thick, with the blade inserted midway between the sides of the stock. The stock is made thicker than the blade so that its face may be applied to the edge of the wood and the steel blade may be laid on the surface to be marked. Usually the blade is provided with a scale of inches divided into eighths.

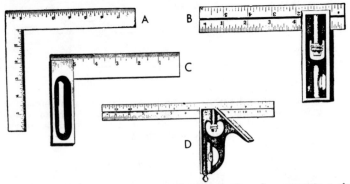

Fig. 5. Various types of squares. In the illustration, A represents a steel square; B, a double try square; C, a try square; D, a combination square. This last square consists of a graduated steel rule with an accurately machined head. The two edges of the head provide for measurements of 45° and 90°.

Miter and Combined Try-and-Miter Squares

The term "miter," strictly speaking, signifies any angle except a right angle, but, as applied to squares, it means an angle of 45°.

In the miter square, the blade (as in the try square) is permanently set but at an angle of 45° with the stock, as shown in Fig. 6.

A try square may be made into a combined try-and-miter square when the end of the stock to which the blade is fastened is faced off at 45°, as along the line MS in Fig. 7. When the 45° face (MS) of the stock is placed against the edge of a board, the blade will be at an angle of 45° with the edge of the board, as in Fig. 8.

An improved form of the combined try-and-miter square is shown in Fig. 9. Because of the longer face (LF) as compared with the short face (MS) in Fig. 7, the blade describes an angle of 45° with greater precision. Its worst disadvantage is that it is awkward to carry because of its irregular shape. However, its precision greatly outweighs its disadvantage.

A square having a blade not exactly at the intended angle is said to be out of true, or simply "out," and good work cannot

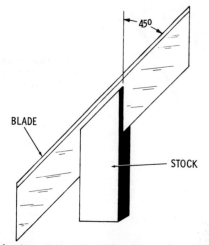

Fig. 6. A typical miter square; it differs from the ordinary try square in that the blade is set at an angle of 45° with the stock, and the stock is attached to the blade midway between its ends.

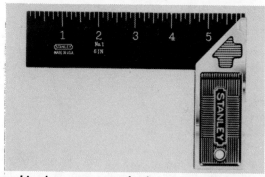

Fig. 7. A combined try square and miter square. Because of its short 45° face (MS), it is not as accurate as the miter square, but it answers the purpose for ordinary marking and the necessity for extra tools.

be done with a square in this condition. A square should be tested and, if found to be out, should be returned to manufacturer.

The method of testing the square is shown in Fig. 10. This test should be made not only at the time of purchase but frequently afterwards, because the tool may become imperfect from a fall or rough handling.

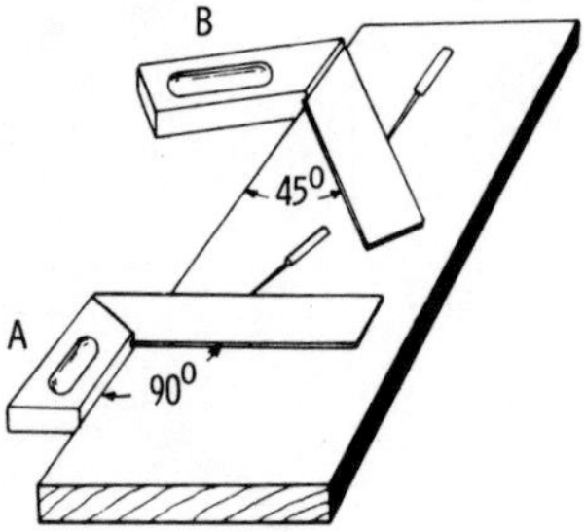

Fig. 8. The combined try-miter square as used for a 90° marking at A and a 45° marking at B.

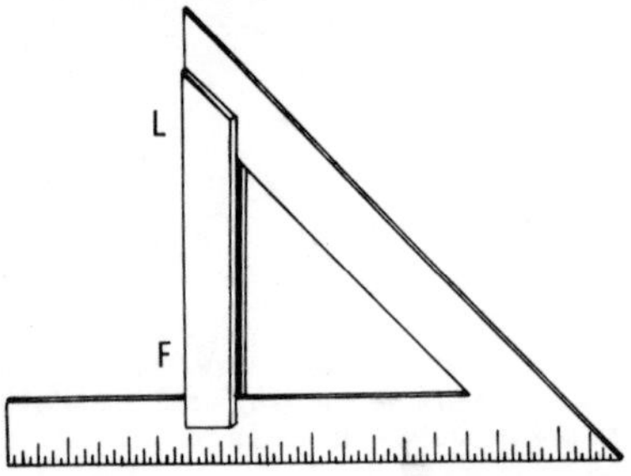

Fig. 9. An improved form of the combined try-miter square.

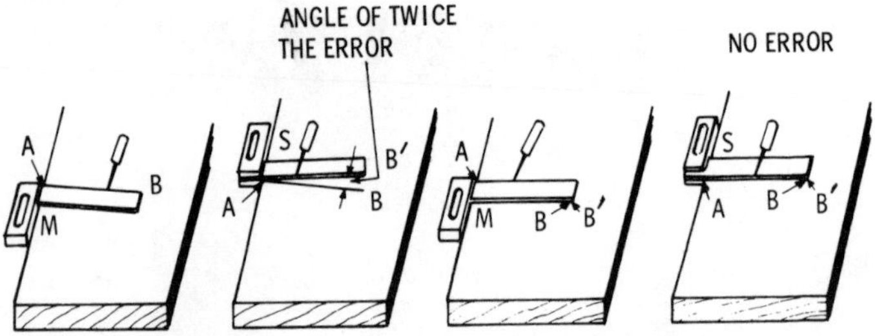

Fig. 10. The method of testing a try square. If the square is "out" (angle not 90°), scribed lines AB and AB' for positions M and S of the square (left side) will not coincide. Angle BAB' is twice the angle of error. If the square is perfect, lines AB and AB' for positions M and S will coincide (right side).

Under no circumstances should initials or other markings be stamped on the brass face of the ordinary try square, because the burrs which project from bending the brass face will throw the square out of truth; for this reason, manufacturers will not take back a square with any marks stamped on the brass face.

Framing or "Steel" Square

The ridiculousness of calling this type of square a steel square is evident from the fact that all types of squares may be

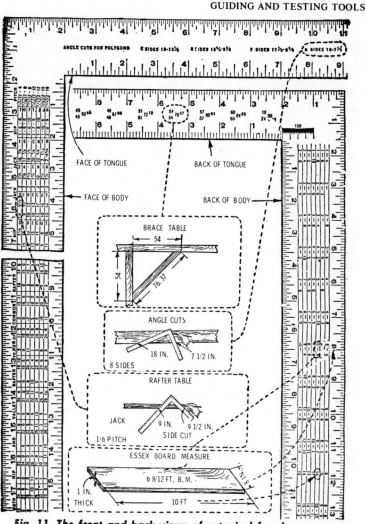

Fig. 11. The front and back views of a typical framing square.

obtained that are made entirely of steel. It is properly called a framing square because with its framing table and various other scales, it is adapted especially for use in house framing, although its range of usefulness makes it valuable to any woodworker. Its general appearance is shown in Fig. 11.

89

The framing square consists of two essential parts—the tongue and the body, or blade. The tongue is the shorter, narrower part, and the body is the longer, wider part. The point at which the tongue and the body meet on the outside edge is called the heel.

There are several grades of squares known as polished, nickled, blued, and royal copper. The blued square with figures and scales in white is perhaps the most desirable. A size that is widely used has an 18-inch body and a 12-inch tongue, but there are many uses which require the largest size whose body measures 24 by 2 inches and whose tongue measures 16 or 18 by 1½ inches.

The feature that makes this square so valuable a tool is its numerous scales and tables. These are:

1. Rafter or framing table.
2. Essex table.
3. Brace table.
4. Octagon scale.
5. Hundredths scale.
6. Inch scale.
7. Diagonal scale.

Rafter or Framing Table—This is always found on the body of the square. It is used for determining the length of common valley hip and jack rafters and the angles at which they must be cut to fit at the ridge and plate. This table appears as a column six lines deep under each inch graduation from 2 to 18 inches, as seen in Fig. 12A, which shows only the 12-inch section of this table; at the left of the table will be found letters indicating the application of the figures given. Multiplication and angle symbols are applied to this table to prevent errors in laying out angles for cuts.

Essex Table—This is always found on the body of the square, as shown in Fig. 12B. This table gives the board measure in feet and twelfths of a foot of boards 1 inch thick of usual lengths and widths. On certain squares, it consists of a table eight lines deep under each graduation, as seen in the figures which represent the 12-inch section of this table.

Brace Table—This table is found on the tongue of the square, a section of which is shown in Fig. 12C. The table gives the length of the brace to be used where the rise and run are from 24 to 60 inches and are equal.

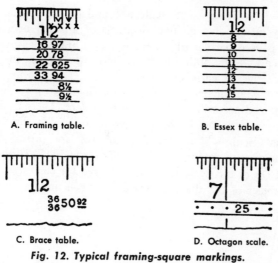

A. Framing table.

B. Essex table.

C. Brace table.

D. Octagon scale.

Fig. 12. Typical framing-square markings.

Octagon Scale—This scale is located on the tongue of the square, as shown Fig. 12D, and is used for laying out a figure with eight sides on a square piece of timber. On this scale, the graduations are represented by 65 dots located $5/24$ of an inch apart.

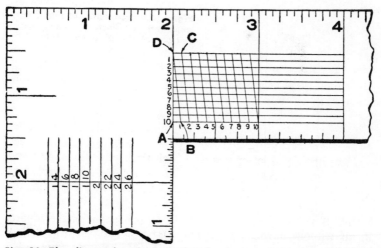

Fig. 13. The diagonal scale on a framing square is used to mark off hundredths of an inch with dividers.

Hundredths Scale—This scale is found on the tongue of the square; by means of a divider, decimals of an inch may be obtained. It is used particularly in reference to brace measure.

Inch Scales—On both the body and the tongue, there are (along the edges) scales of inches graduated in $\frac{1}{32}$, $\frac{1}{16}$, $\frac{1}{12}$, $\frac{1}{10}$, $\frac{1}{8}$, and $\frac{1}{4}$ of an inch. Various combinations of graduations can be obtained according to the type of square. These scales are used in measuring and laying out work to precise dimensions.

Diagonal Scale—Many framing squares are provided with what is known as a diagonal scale, as shown in Fig. 13; one division (ABCD) of this scale is shown enlarged for clearness in Fig. 14. The object of the diagonal scale is to give minute measure-

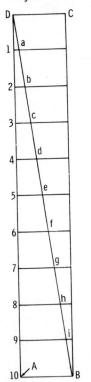

Fig. 14. Section ABCD of Fig. 13, enlarged to illustrate the principle of the diagonal scale.

ments without having the graduations close together where they would be hard to read. In construction of the scale (Fig. 14), the short distance AB is $\frac{1}{10}$ of an inch. Evidently, to divide

AB into ten equal parts would bring the divisions so close together that the scale would be difficult to read. Therefore, if AB is divided into ten parts, and the diagonal BD is drawn, the intercepts 1a, 2b, 3c, etc., drawn through 1, 2, 3, etc., parallel to AB, will divide AB into $\frac{1}{10}$, $\frac{2}{10}$, $\frac{3}{10}$, etc. of an inch. Thus, if a distance of $\frac{3}{10}$ AB is required, it may be picked off by placing one leg of the dividers at 3 and the other leg at c, thereby producing $3c = \frac{3}{10}$ AB.

Because of the importance of the framing square and the many problems to be solved with it, the applications of the square are given at length in a later chapter.

Combination Square

This tool (Fig. 15), as its name indicates, can be used for the same purposes as an ordinary try square, but it differs from the try square in that the head can be made to slide along the blade and clamp at any desired place; combined with the square,

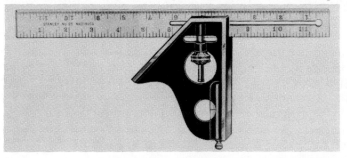

Fig. 15. A typical combination square with a grooved blade, level, and centering attachments.

it is also a level and a miter. The sliding of the head is accomplished by means of a central groove in which a guide travels in the head of the square. This permits the scale to be pulled out and used simply as a rule. It is frequently desired to vary the length of the try-square blade; this is readily accomplished with the combination square. It is also convenient to square a piece of wood with a surface and at the same time tell whether one or the other is level, or plumb. The spirit level in the head of the square permits this to be done without the use of a separate level. The head of the square may also be used as a simple level.

Because the scale may be moved in the head, the combination square makes a good marking gauge by setting the scale at the proper position and clamping it there. The entire combination square may then be slid along as with an ordinary gauge. As a further convenience, a scriber is held frictionally in the head by a small brass bushing. The scriber head projects from the bottom of the square stock in a convenient place to be withdrawn quickly.

In laying out, the combination square may be used to scribe lines at miter angles as well as at right angles, since one edge of the square head is at 45°. Where micrometer accuracy is not essential, the blade of the combination square may be set at any desired position, and the square may then be used as a depth gauge to measure in mortises, or the end of the scale may be set flush with the edge of the square and used as a height gauge.

The head may be unclamped and entirely removed from the scale, and a center head can then be substituted so that the same tool can quickly be used to find the centers of shafting and other cylindrical pieces. In the best construction, the blade is hardened to prevent the corner from wearing round and destroying the graduations, thus keeping the scale accurate at all times. This combination square combining as it does a rule, square, miter, depth gauge, height gauge, level, and center head permits more rapid work on the part of the carpenter, saves littering the bench with a number of tools each of which is necessary but which may be used only rarely, and tends toward the goal for which all carpenters are striving—greater efficiency. Some of the uses for the combination square are illustrated in Fig. 16.

SLIDING "T" BEVEL

A bevel is virtually a try square with a sliding adjustable blade that can be set at any angle to the stock. In construction, the stock may be of wood or steel; when the stock is made of wood, it normally has brass mountings at each end and is sometimes concaved along its length. The blade is of steel with parallel sides, and its end is at an angle of 45° with the sides, as shown in Fig. 17. The blade is slotted, thereby allowing linear adjustment and the insertion of a pivot, or screw pin, which is located at the end of the stock. After the blade has been adjusted to any particular angle, it is secured in position by tightening the screw lever on

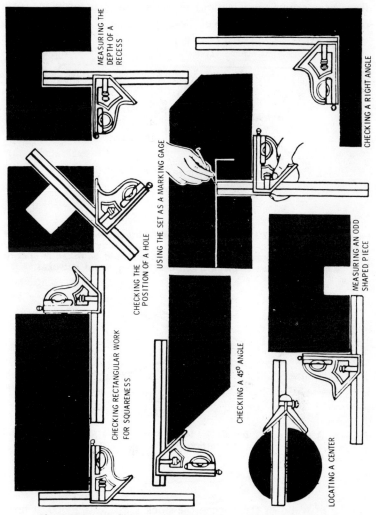

Fig. 16. Some of the many uses of the combination square.

the pivot; this action compresses the sides of the slotted stock together, thus firmly gripping the blade. Fig. 18 illustrates how to set the blade angle.

When selecting a bevel, care should be taken to see that the edges are parallel and that the pivot screw, when tightened, holds the blade firmly without bending it. In the line of special bevels, there are various modifications of the standard or ordinary form of bevel just described. Two of these are shown in Figs. 19 and 20.

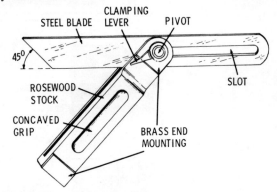

Fig. 17. A sliding "T" bevel with a steel blade, rosewood stock, and brass end mountings. Since the size of a bevel may be expressed by the length of either its stock or its blade, care should be taken to specify which dimension is given when ordering to avoid mistakes.

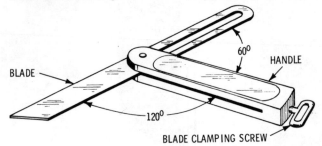

Fig. 18. A sliding "T" bevel. A tool of this type is used to mark and test cutting angles.

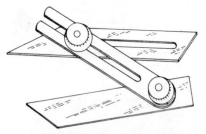

Fig. 19. A double-slot steel bevel. As shown, both the stock and the blade are slotted, thus permitting adjustments that cannot be obtained with a common bevel.

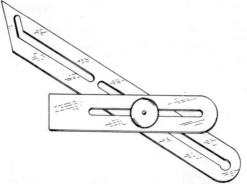

Fig. 20. A typical combination bevel. One leg is pivoted to a straight-edge, as shown, so that it can swing over the stock and be clamped at any angle. The slotted auxiliary blade may be slipped on the split blade and clamped at any desired angle to be used in conjunction with the stock for laying out work.

MITER BOX

This device is used to guide the saw in cutting work to form miters, and it consists of a trough formed by a bottom and two side pieces of wood screwed together, with a saw cut through the sides at angles of 45° and 90°, as shown in Fig. 21. Note that there are two 45° cuts; these are for cutting right and left miters.

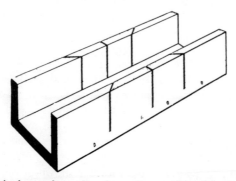

Fig. 21. A typical wooden miter box. A nonadjustable 45° and 90° miter box, as shown, may easily be constructed from three pieces of hardwood screwed together as indicated.

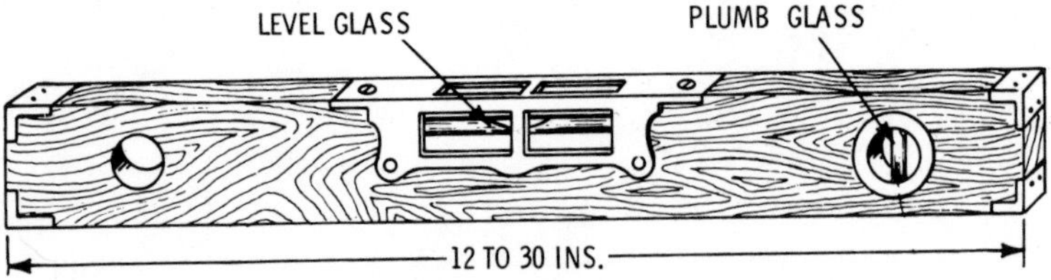

LEVEL GLASS   PLUMB GLASS

12 TO 30 INS.

Fig. 22. A typical wooden spirit level with a horizontal and a vertical tube.

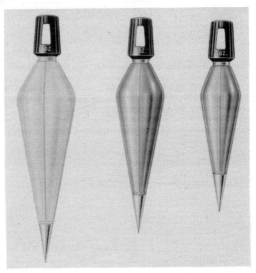

Fig. 23. The solid plumb bob.

LEVEL

This tool is used for both guiding and testing—to guide in bringing the work to a horizontal or vertical position, and to test the accuracy of completed construction. It consists of a long rectangular body of wood or metal that is cut away on its side and near the end to receive glass tubes, which are almost entirely filled with a nonfreezing liquid that leaves a small bubble free to move as the level is moved. A typical level is shown in Fig. 22.

The side and end tubes are at right angles, so that when the bubble of the side tube is in the center of the tube, the level is

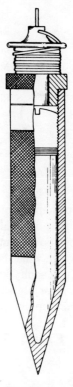

Fig. 24. A mercury-filled plumb bob. This type of plumb bob has the advantage of great weight in proportion to its surface area, and it is considerably better suited than the one shown in Fig. 23 when working outside in a wind.

horizontal; when the bubble of the end tube is in the center, the level is vertical. By holding the level on a surface supposed to be horizontal or vertical, it may be ascertained whether the assumption is correct or not.

PLUMB BOB

The word "plumb" means perpendicular to the plane of the horizon, and since the plane of the horizon is perpendicular to the direction of gravity at any given point, the force due to gravity is utilized to obtain a vertical line in the device known as a plumb bob.

This tool consists of a pointed weight attached to a string. When the weight is suspended by the string and allowed to come to rest,

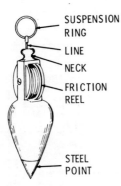

SUSPENSION
RING

LINE

NECK

FRICTION
REEL

STEEL
POINT

*Fig. 25. An adjustable
plumb bob.*

as in Fig. 23, the string will be plumb (vertical). The ordinary top-shaped solid plumb bob is objectionable because of a too-blunt point and not enough weight. For outside work, the second objection is important, since when the plumb bob is used with a strong wind blowing, the excess surface presented to the wind will magnify the error, as shown in Fig. 23. To reduce the surface for a given weight, the bob is bored and filled with mercury. This type of plumb bob is shown in Fig. 24. An adjustable bob with a self-contained reel on which the string is wound is shown in Fig. 25. The convenience of this arrangement is apparent.

SUMMARY

It requires considerable skill to make a good joint, because the parts depend on accuracy in measurement and precision in angles and cutting. The material must be made straight and cut to the proper length. In order to insure accuracy, various tools of guidance and direction are used.

A tool that is used to guide the pencil or scriber when marking a straight line and testing a straight surface is the straightedge. This tool may be made either of wood or steel in various lengths. The try square is used to check for right angle cuts on any straightedged material. There are various types of try squares, such as double try squares, combination try squares, and try-and-miter squares.

A framing square (sometimes called steel square) is a square with framing tables and various other scales. It is adapted especially for use in house framing. The features that makes this square such a valuable tool is the table of rafter measurements for common, valley, hip, and jack rafters, and the angles at which they must be cut to fit the ridge and plate.

The miter box is a tool used to guide the saw in cutting material, generally at 45° and 90° angles. A miter box, in most cases, has two 45° angles; these are for cutting right and left miters.

The level is a tool used to bring the work to either a true horizontal or vertical position. By holding the level on a surface supposed to be horizontal or vertical, it may be checked for accuracy.

REVIEW QUESTIONS

1. What is a straightedge?
2. Explain the use of the plumb bob and the level.
3. Name a few of the common types of squares.
4. What is a shooting board?
5. Why does a miter box have two 45° angles?

CHAPTER 8

Marking Tools

In good carpentry and joinery, a great deal depends on the accuracy achieved in laying out the work. The term "laying out" means the operation of marking the work with a tool, such as a pencil or scriber, so that the various centers and working lines will be set off in their proper relation. These lines are followed by the carpenter in cutting and other tooling operations necessary to bring the work to its final form.

In laying out, the guiding tools just described in Chapter 7 are used to guide the pencil or scriber; the measurements are made by the aid of the measuring devices described in Chapter 9.

According to the degree of precision desired in laying out, the proper marker to use is

1. For extremely rough work—the chalk box and reel or the carpenter's pencil with rectangular lead.
2. For rough work—the lead pencil with round lead.
3. For semirough work—the scratch awl.
4. For precision work—the scriber or knife.

For efficiency, the carpenter must use judgment to work with the proper degree of precision. Thus, it would be ridiculous to use a machine-hardened steel scriber with a needle point to mark off rafters, or to use a carpenter's pencil with "an acre of soft lead" on the point to lay out a fine dovetail joint. Simply exercise common sense.

CHALK BOX AND LINE

The special device, which is shown in Fig. 1 is to mark a long straight line between two points that are too far apart to permit the use of a square or straightedge.

The chalk box is usually constructed of aluminum or plastic. Inside the chalk box is a reel that is fitted with a lightweight string or cord. In order for the lines to leave a mark, powdered chalk is placed inside the box. The line can then be stretched between two points, when the string is taut, it is pulled up and released, thus leaving a chalked line on the surface of the work. Note the right way and the wrong way to use the chalk line, as shown in Fig. 2.

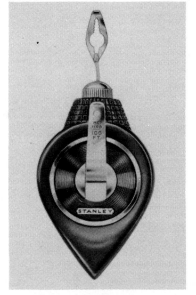

Fig. 1. A chalk box with 100 feet of string.

CARPENTER'S PENCIL

The conventional carpenter's pencil has a lead with a rectangular cross section which is considerably larger than an ordinary pencil. The object of making the lead in this shape is to permit its use on rough lumber without too frequent sharpening and to give a well-defined, plainly visible line. Because of the width of the line, the carpenter's pencil is not intended for fine work but is used principally for marking boards, etc., that are to be sawed.

Fig. 3 shows the general appearance of the carpenter's pencil and the shape of the lead. When marking with the carpenter's pen-

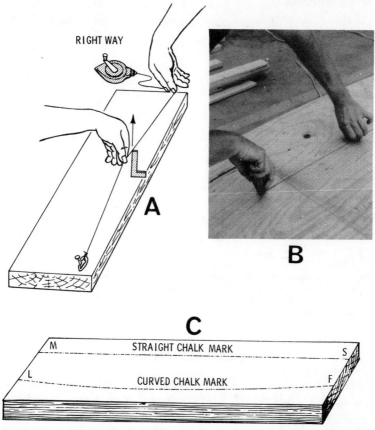

Fig. 2. *The right way to use the chalk box. When pulling up the line, always do so in a direction that is at right angles with the board. If the chalk line is pulled straight up, as in A, a straight chalk mark MS will be obtained; if the line is pulled up to one side, a curved line LF will be produced.*

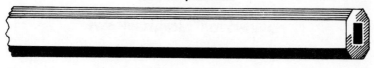

Fig. 3. *A typical carpenter's pencil.*

cil, the mark must be made in the direction of the long axis of the lead, as shown in Fig. 4B, and not as in Fig. 4C. The

105

proper method of sharpening the pencil, as shown in Fig. 4A, should be noted.

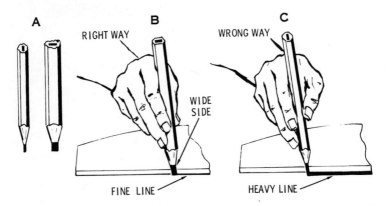

Fig. 4. The right and wrong ways to use the carpenter's pencil; A, side and end views of a carpenter's pencil; B, a fine line is obtained with the long side of the lead turned in the direction of the straightedge; C, a wide, undefined line is produced when the pencil is used in this position.

ORDINARY PENCIL

This form of pencil, with its cylindrical lead, is familiar to all and needs no description. Since the lead is smaller than that of the carpenter's pencil, it produces a finer marking line. It is used on smooth surfaces where more accurate marking is required than can be obtained with the carpenter's pencil. When using, the best results are obtained by twisting the pencil while drawing the lines so as to retain the conical shape given the lead in sharpening.

MARKING OR "SCRATCH" AWL

This tool consists of a short piece of round steel that is pointed at one end with the other end permanently fixed in a convenient handle, as shown in Fig. 5. A scratch awl is used in laying out fine work where a lead pencil mark would be too coarse for the required degree of precision.

106

Fig. 5. An ordinary scratch awl with a forged blade and a hardwood handle.

SCRIBER

This is a tool of extreme precision, and, while intended especially for machinists, it should be in the tool kit of carpenters and all mechanics who make any claim to being skilled in their occupation.

A scriber is a hardened steel tool with a sharp point designed to mark extremely fine lines. The most convenient form of scriber is the pocket, or telescoping, type, shown in Fig. 6; the construction renders it safe to carry in the pocket.

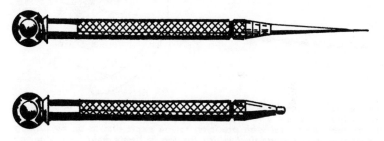

Fig. 6. A telescoping scriber in the open and closed positions.

COMPASS AND DIVIDERS

The compass is an instrument used for describing circles or arcs by scribing. It consists of two pointed legs that are hinged firmly by a rivet so as to remain set in any position by the friction of the hinged joint. The usual form of carpenter's compass is shown in Fig. 7, and it should not be used in place of dividers for dividing an arc or line into a number of equal divisons, because it is not a tool of precision.

107

Fig. 7. A typical compass.

The difference between dividers and compasses is that the dividers are provided with a quadrantal wing projecting from one of the two hinged legs through a slot in the other leg. A set-screw on the slotted leg enables the instrument to be securely locked to the approximate dimension and adjusted with precision to the exact dimension by a screw at the other end of the wing. A spring pressing against the wing holds the leg firmly against the screw. Its general appearance is shown in Fig. 8. Because of the wing, the tool is frequently called winged dividers.

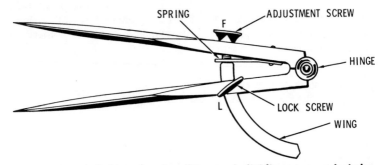

Fig. 8. Winged dividers for describing and dividing arcs and circles. When the dividers are locked in the approximate setting by lock screw L, the tool can be set with precision to the exact dimension by turning adjustment screw F, against which the leg is firmly held by the spring to prevent any lost motion.

SUMMARY

Accuracy in carpentry work depends on the correct use of good tools. In layout work, the guiding tools are used to guide a pencil or scriber.

The conventional carpenter's pencil has a rectangular lead which is considerably larger than that in an ordinary pencil.

With this design, the pencil can be used frequently without sharpening the lead.

The scratch awl is a short piece of round steel that is pointed at one end with the other end permanently fixed in a convenient handle. A scratch awl is used in laying out fine work where a pencil would be too coarse for the required precision.

The compass or divider is an instrument used for describing circles or arcs. It is designed with two pointed legs that are hinged at one end.

REVIEW QUESTIONS

1. What is the difference between a divider and a compass?
2. Why is a carpenter pencil lead rectangular?
3. What is the purpose of the scratch awl?
4. What marking tool would be used for precision work?
5. What is a chalk line and how is it used?

CHAPTER 9

Measuring Tools

In laying out work, after having scribed a line with one of the marking tools just described, aided by a guiding tool, the next step is usually to measure off on the scribed line by some given distance. This is done with a suitable measuring tool. There are many kinds of measuring tools, known as rules, of which the following are of particular interest to the carpenter.

CARPENTER'S TWO-FOOT RULE

This is the most familiar form of rule and is usually made fourfold, that is, with three hinges spaced 6 inches apart and so arranged that it can be folded up, as shown in Fig. 1. It is usually made of boxwood with different grades of metal mountings and graduations. The more expensive rules are normally divided into

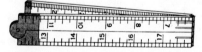

Fig. 1. A typical 2-foot four-fold boxwood rule with round joints and middle plates. The rule is graduated in 8ths and 16ths.

16ths, 12ths, 10ths, and 8ths. When selecting a rule, buy only the best quality available; the kind of joint provided should also be noted. These are:

1. Round joint.
2. Square joint.
3. Arch joint.
4. Double arch joint.

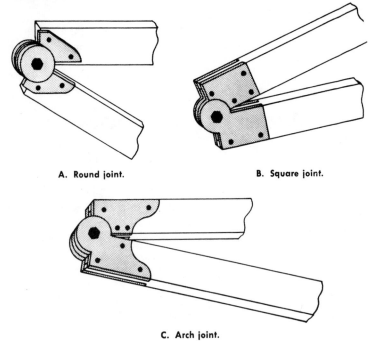

A. Round joint.   B. Square joint.

C. Arch joint.

Fig. 2. Two-foot folding rule joints.

Round Joint

The round joint, as shown in Fig. 2A, is the cheapest (and weakest), since it has only one flange or wing embedded in each leg of the rule, and the leg and wing are pinned together.

Square Joint

A better construction is the square joint, shown in Fig. 2B; it has two wings to each leg—one on each outside face of the wood. The two wings are held together by rivets which go through all three, thereby providing unusual strength.

Arch Joint

The arch joint, Fig. 2C, is a still better construction, since the wings are larger and cover more surface of the wood, thus adding to the life of the rule.

Double Arch Joint

The best construction is the double arch joint; this is substantially of the same construction as the arch joint but is repeated at the folding point, which again adds to the strength of the rule.

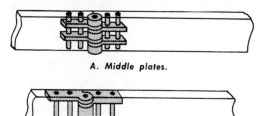

A. Middle plates.

B. Edge plates.

Fig. 3. Two-foot folding rule plates.

Folding Joint

The folding joint is made in two styles, one with middle plates and one with edge plates. The middle-plate construction (Fig. 3A) is that in which the plates are let in near the center of the wood and pinned. The edge-plate construction (Fig. 3B) is that in which the plates are fastened to the outer edges of the wood by rivets which go through both the wood and the plates and hold the three firmly together. Edge-plate joints are stronger than middle-plate joints.

Characteristics of Rules

A full-bound rule is one with a brass binding extending along both inside and outside edges of each leg. A half-bound rule is one with a brass binding extending only along the outside edges of the legs. Bitted rules have a brass plate inserted on the edge of the rule to protect the wood from the closing pin.

In the American marking system, the numbers run from right to left, as in Fig. 4A, and in the English marking system, they run from left to right, as in Fig. 4B. The two-foot rule may be used as a protractor with the aid of Table 1.

Fine quality rules are provided with architect's scales, such as ¼ in.= 1 ft.; ½ in.=1 ft., etc., for directly measuring actual distances on a small size drawing.

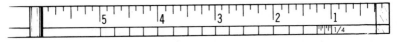

A. American system—graduated from right to left.

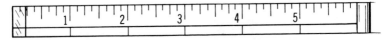

B. English system—graduated from left to right.

Fig. 4. Two marking systems for the carpenter's 2-foot folding rule.

Table 1. Angles and Openings

Ang. °	Dis. in.	Ang. °	Dis. in.	Ang. °	Dis. in.	Ang. °	Dis. in.	Ang. °	Dis. in.	Ang. °	Dis. in.
1	0.20	16	3.34	31	6.41	46	9.38	61	12.18	76	14.78
2	0.42	17	3.55	32	6.62	47	9.57	62	12.36	77	14.94
3	0.63	18	3.75	33	6.82	48	9.76	63	12.54	78	15.11
4	0.84	19	3.96	34	7.02	49	9.95	64	12.72	70	15.27
5	1.05	20	4.17	35	7.22	50	10.14	65	12.90	80	15.43
6	1.26	21	4.37	36	7.42	51	10.33	66	13.07	81	15.59
7	1.47	22	4.58	37	7.61	52	10.52	67	13.25	82	15.75
8	1.67	23	4.78	38	7.81	53	10.71	68	13.42	83	15.90
9	1.88	24	4.99	39	8.01	54	10.90	69	13.59	84	16.06
10	2.09	25	5.19	40	8.20	55	11.08	70	13.77	85	16.21
11	2.30	26	5.40	41	8.40	56	11.27	71	13.94	86	16.37
12	2.51	27	5.60	42	8.60	57	11.45	72	14.11	87	16.52
13	2.72	28	5.81	43	8.80	58	11.64	73	14.28	88	16.67
14	2.92	29	6.01	44	8.99	59	11.82	74	14.44	89	16.82
15	3.13	30	6.21	45	9.18	60	12.00	75	14.61	90	16.97

VARIOUS FOLDING RULES

In addition to the standard rule just described, rules are regularly made of similar construction in lengths from 6 inches to 4 feet, as follows:

Twofold, 6 inches and 2 feet.
Fourfold, 1 foot, 3 feet, and 4 feet.

For bench work, where compactness is not so important as for outdoor work, the length between the joints may be longer, so that in many cases, the twofold 2-foot rule and the fourfold rules of 3- and 4-foot lengths are more desirable than with a greater number of joints.

RULES WITH ATTACHMENTS

To increase the utility of carpenters' rules, they may be obtained with various attached devices, such as calipers, levels, and protractors. Caliper rules are regularly made with the caliper for the left hand, as shown in Fig. 5, but may be obtained for the right hand.

The level device is fitted as shown in Fig. 6, and is attached to the center section of a threefold rule; each outer section is hollowed to fit the level tube, thereby leaving the glass visible when the rule is closed.

The protractor is attached to the rule as illustrated in Fig. 7. With the aid of the protractor, any size angle may be measured.

Fig. 5. A 1-foot two-fold boxwood caliper rule with square joints. The rule is graduated in 8ths, 10ths, 12ths, 16ths, and 32nds.

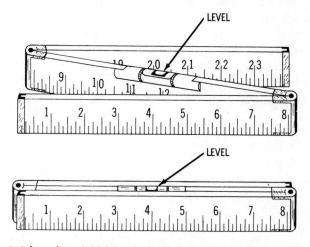

Fig. 6. A 2-foot three-fold boxwood combination rule and level, graduated in 8ths and 16ths.

The level also permits the rule to be employed for leveling, although it is too short for any real accuracy.

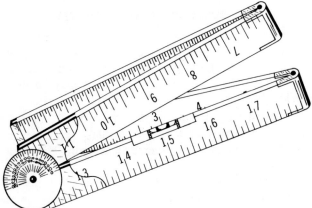

Fig. 7. A 2-foot four-fold boxwood combination rule, level, and protractor, graduated in 8ths and 16ths.

BOARD MEASURE

Lumber is estimated according to a system of arbitrary surface measure known as board measure (B.M.). Any board less than 1 inch in thickness is considered as being 1 inch thick; any board over 1 inch in thickness is measured in inches and fractions of an inch. Thus, a ½-inch board is considered as being 1 inch thick, and a 1½-inch board is considered as being 1½ inches thick. To find the number of feet board measure, multiply the length in feet by the width in feet of the board, and multiply this product by 1 for boards of 1 inch or less than 1 inch thick and by the thickness in inches and fractions of an inch for boards over 1 inch thick.

Example—How many feet board measure are there in a board 12 feet long by 18 inches wide by 1¾ inches thick?

$$18 \text{ in. width} = (18 \div 12) \text{ ft.} = 1.5 \text{ ft.}$$

$$12 \times 1.5 \times 1.75 = 31.5 \text{ ft. B.M.}$$

Example—How many feet board measure would there be in the board of the previous example if the board were only ½ inch thick?

$$(12 \times 1.5) \times 1 = 18 \text{ ft. B.M.}$$

LUMBER SCALES

When estimating lumber, a great deal of time can be saved by using a lumber scale by means of which the board feet measure may be read off directly. This scale gives an approximate result. When using the scale, it is customary to read to the nearest figure and when there is no difference to alternate between the lower and higher figures on different boards. Fig. 8 shows a board scale graduated for boards of 12-, 14-, and 16-foot lengths and the method of using the scale.

There are many types of lumber scales in use in various sections of the United States and Canada. Table 2 compares the measurements of a 16-foot log as given by the principal log scales.

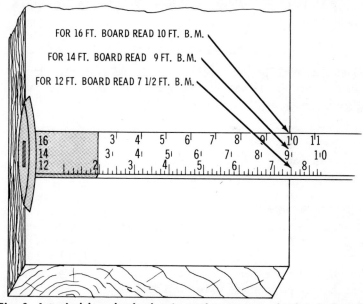

Fig. 8. A typical board rule that is used to measure lumber in board-measure units. To use the rule, place the head of the rule against one edge of the board, and read the figure nearest the other edge (width) of the board in the same line of figures on which the length is found. This reading will give the number of feet board measure for the piece of lumber being measured.

Biltmore or Forest Cruiser Stick

This stick gives the diameter and height of standing trees and was designed for the U.S. Forest Service in reconnaissance estimates of timber tracts. In addition to diameter and height scales, it bears a table of decimal C log scale values for 16-foot logs and a tier of regular inch markings: $37^{13}/_{16}$ inches long, ½ inch thick, 1 inch wide with one sloping face. Both ends are fitted with a heavy cast brass ferrule.

Spring Steel Board Rule

This rule is made of tempered spring steel so that it will bend to the board and, when released, will return exactly straight. It is provided with a wood handle and a leather slide for handling the rule at any part of the blade. Types—3-tier, $3\frac{1}{2}$-foot inspectors' rule; 3-tier, 3-foot board rule; 3-tier, $2\frac{1}{2}$-foot sorting rule. Markings—all three types are marked on one side to measure 8, 10, and 18 feet; the opposite side is marked 12, 14, 16, or 18, 20, and 22 feet.

MARKING GAUGES

Tools of this type are used to mark a piece of wood that is to be sawed or otherwise tooled. There are several types of marking gauges, such as

1. Single bar.
2. Double bar.
3. Single bar with slide.
4. Butt.

Single-Bar Gauge

This is used for making a single mark, such as for sawing. It consists of a bar with a scriber, or pin, at one end, provided with a scale that is graduated in inches and sixteenths. The bar passes through a movable head that may be clamped at any distance from the scriber point, as shown in Fig. 9.

Double-Bar Gauge

This type of gauge is designed especially for mortise marking. There are two independent bars working in the same head. One pin

118

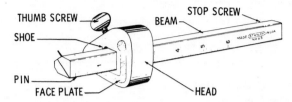

Fig. 9. A single-bar marking gauge. It is fitted with a head, faceplate, and thumbscrew and is provided with a scale that is graduated in inches and 16ths.

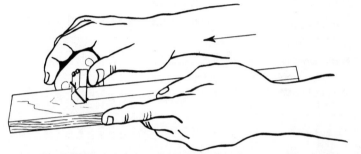

Fig. 10. The method of using a marking gauge. When setting the gauge, use a rule, unless it is certain that the scriber point is located accurately with the graduations on the bar. When marking, the gauge should be held as indicated; the face of the head is pressed against the edge of the board. Care must be taken to keep it true with the edge so that the bar will be at right angles with the edge, and the line scriber will be at the correct distance from the edge. The line is usually scribed by pushing the gauge away from the worker. Always work from the face side, as shown.

is affixed to each bar. After setting the bars for the proper marking of the mortise, one side is marked with one bar, and the gauge is then turned over for marking the other side. The construction is shown in Fig. 11.

Slide Gauge

One objection to the double-bar gauge is that two operations are required that can both be performed with a slide gauge in one operation. As shown in Fig. 12, the underside of the bar is provided with a flush slide having a scriber B at the end of the slide, with another scriber A at the end of the bar. These two scribers, when set to the required distances from the head, mark both sides of the tenon or mortise to the size required with just one

119

Table 2. Comparison

	Dayle Scale																				
Combined Doyle and Scribner Scale	16	25	36	49	64	81	100	121	144	169	196	225	256	289	324	359	400	441	484	530	576
Spaulding's Pacific Coast Scale			50	63	77	94	114	137	161	188	216	245	276	308	341	376	412	449	488	528	569
Decimal Scale	3	4	6	7	8	10	11	14	16	18	21	24	28	30	33	38	40	46	50	55	58
Northwestern Scale	33	45	61	70	77	97	117	144	170	188	206	226	248	285	324	357	392	421	450	520	536
St. Louis Hardwood Scale	33	46	59	72	85	100	116	133	150	172	192	213	237	261	284	312	341	369	400	432	464
Cumberland River Scale			47	57	68	80	93	107	121	137	153	171	190	209	229	250	281	296	320	345	372
Dusenberry Scale			40	54	68	80	100	117	136	157	180	204	229	256	285	315	346	379	414	450	487
Two-thirds Scale	36	48	58	70	85	100	116	133	151	171	192	213	237	261	286	313	341	370	400	432	464
Baxter Scale	34	44	56	69	84	100	117	136	156	177	200	224	250	277	305	335	366	399	432	468	504
Favorite Scale	22	29	37	48	64	82	98	120	142	166	197	226	248	285	324	357	392	434	476	520	562
Scribner Scale	25	36	49	64	79	97	114	142	159	185	213	240	280	304	334	377	404	459	500	548	582
Doyle's Scale	16	25	36	49	64	81	100	121	144	169	196	225	256	289	324	359	400	441	484	530	576
Diameter in Inches	8	9	10	11	12	13	14	15	16	17	18	19	20	21	22	23	24	25	26	27	28

stroke. On the upper side, there is one scriber C for single marking.

of Log Tables

	Scribner Scale																			
Combined Doyle and Scribner Scale	609	657	710	736	784	800	876	923	1029	1068	1120	1204	1272	1343	1396	1480	1518	1587	1656	1728
Spaulding's Pacific Coast Scale	612	656	701	748	796	845	897	950	1006	1064	1124	1185	1248	1312	1377	1448	1512	1581	1652	1724
Decimal Scale	61	66	71	74	78	80	88	92	103	107	112	120	127	134	140	148	152	159	166	173
Northwestern Scale	584	632	678	725	785	845	882	920	978	1037	1090	1160	1213	1266	1334	1402	1474	1546	1621	1696
St. Louis Hardwood Scale	498	533	568	608	645	685	725	768	810	856	901	947	996	1045						
Cumberland River Scale	399	427	546	485	516	548	581	614	649	685	721	759	798	835	877	918	960	1003	1048	1092
Dusenberry Scale	526	567	609	652	697	744	792	841	892	945	999	1054	1111	1170	1227	1300	1350	1410	1470	1580
Two-thirds Scale	498	533	569	606	648	685	725	768	811	855	901	944	995	1045	1095	1147	1200	1253	1309	1365
Baxter Scale	543	582	623	665	709	754	800	848	897	946	999	1052	1107	1163	1221	1280	1340	1401	1464	1529
Favorite Scale	596	632	678	725	785	845	882	920	978	1037	1098	1160	1213	1266	1334	1402	1474	1546	1621	1696
Scribner Scale	609	657	710	736	784	800	876	923	1029	1068	1120	1204	1272	1343	1396	1480	1518	1587	1656	1728
Doyle's Scale	625	676	729	784	841	900	961	1024	1089	1156	1225	1296	1369	1444	1521	1600	1681	1764	1849	1936
Diameter in Inches	29	30	31	32	33	34	35	36	37	38	39	40	41	42	43	44	45	46	47	48

Butt Gauge

When hanging doors, there are three measurements to be marked:

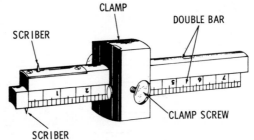

Fig. 11. A double-bar marking gauge. It is used for marking a given distance between two parallel lines and a given distance from the edge of a board. As shown in the illustration, each bar has a setscrew for clamping it in any desired position.

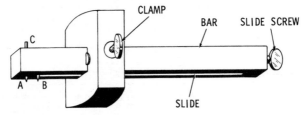

Fig. 12. A slide marking gauge. The bar has a scriber (C) on the upper side for single marking and a scriber (A) on the lower side which, with the scriber (B) on the slide, works flush in the bar. The distance between scriber points A and B is regulated by the slide screw at the end of the bar.

1. The location of the butt on the casing
2. The location of the butt on the door.
3. The thickness of the butt on the casing.

A butt gauge is a type of gauge having three cutters, which are purposely arranged so that no change of setting is necessary when hanging several doors. In reality, these tools comprise rabbet gauges, marking gauges, and mortise gauges of a scope sufficient for all door trim, including lock plates, strike plates, etc.

Fig. 14 shows a typical butt gauge. The cutters are mounted on the same bar and are set by one adjustment with the proper allowance for clearance. When casings have a nailed-on strike instead of being rabbeted, a marking gauge that will work on a ledge as narrow as ⅛ inch is required; in this case, the same distance is marked from the edge of the casing and from the edge of the door that is not engaged when closing. Certain gauges can be

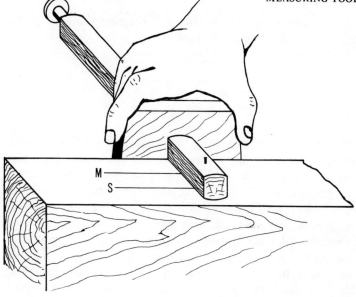

Fig. 13. *The method of using a slide gauge when marking a mortise. Note that the marks (M and S) for the sides of the mortise can both be scribed in one operation.*

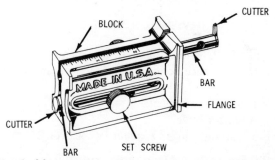

Fig. 14. *A typical butt gauge. It is used to mark the location of the butt on casings and doors. Three separate cutters, one for each dimension, eliminate changing the setting when more than one door is hung. It is also used as a marking and mortising gauge and an inside and outside square for squaring the edge of a butt on the door and jamb; it is graduated in 16ths for 2 inches.*

used on such work; one cutter marks the butt, and one cutter marks its thickness. Other gauges are made so that they can be used as inside or outside squares for squaring the edge of the butt on either the door or the jamb.

SUMMARY

The most common measuring tool known in any type of work is the rule. There are many different types of so-called tools that fall in this category, but in this type of work it is referred to as the carpenter's rule. The most familiar type is the 2-foot folded wooden rule. There are various styles of folded rules, such as round joint, square joint, and arch joint.

Generally, rules are divided into 16ths, 12ths, 10ths, and 8ths. In the American marking system, the numbers read from right to left, and in the English marking system, they read from left to right. Some carpenter's rules are provided with architect's scales, such as ¼ in. = 1 ft.; ½ in. = 1 ft., etc.

Marking gauges are used to mark pieces of wood that are to be sawed. There are several types of marking gauges, such as single-bar, double-bar, single-bar with slider, and butt.

REVIEW QUESTIONS

1. What type of rule is most familiar in carpentry work?
2. What is the difference between the American and English style of numbering?
3. Explain board measurement.
4. What is a lumber scale?
5. What is a marking gauge? Explain how it works.

CHAPTER 10

Holding Tools

An essential part of the shop equipment necessary for good carpentry is the proper assortment of holding tools, since, as must be evident, there are many tooling operations which require that the work be held rigid, even when considerable force is applied, such as in planing and chiseling.

The workbench, considered broadly with its attachments, may be called the main holding tool, and unless this important part of the equipment is constructed amply substantial and rigid, it will be difficult to do good work. The workbench and its attachments have already been described. When installing a bench, it should be properly anchored or fastened to the wall of the building so as to be as rigid as though it were a part of the building.

Holding tools may be generally classed as supporting tools and retaining tools. When marking or sawing, it is usually only necessary to support the work by placing it on the bench or on sawhorses; however, in planing, chiseling, and some nailing operations, the work must not only be supported but held rigidly in position.

"HORSES" OR TRESTLES

These are used in various ways to simply support the work when it is of such large dimensions that the bench cannot conveniently be used, especially for marking and sawing planks. No shop equipment list is complete without a pair of sawhorses. A sawhorse, as usually made, consists of a 3- or 4-foot length of $2'' \times 4''$ or $2'' \times 6''$ stock for the cross beam, with a pair of $1'' \times 3''$ or $1\frac{1}{4}'' \times 4''$ legs at each end, depending on the expected weight of the work.

The height of the sawhorse is usually 2 feet. The general construction of sawhorses is shown in Fig. 1. Note that the legs are

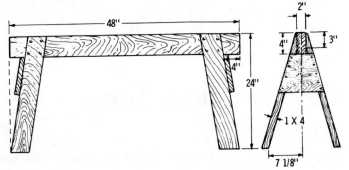

Fig. 1. The side and end views of a typical carpenter's sawhorse whose dimensions are suitable for general use.

inclined outward both lengthwise and crosswise, and a problem arises as to how to determine the length of the legs having this double inclination for a given height of sawhorse; the solution is shown in Fig. 2, and the method of obtaining the angular setting for the bevel to scribe the mortise is illustrated in Fig. 3.

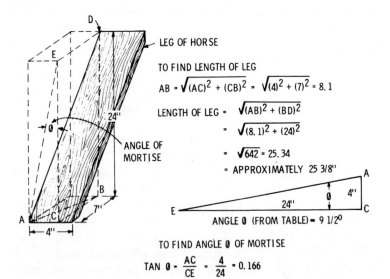

LEG OF HORSE

TO FIND LENGTH OF LEG

$$AB = \sqrt{(AC)^2 + (CB)^2} = \sqrt{(4)^2 + (7)^2} = 8.1$$

LENGTH OF LEG $= \sqrt{(AB)^2 + (BD)^2}$

$$= \sqrt{(8.1)^2 + (24)^2}$$

$$= \sqrt{642} = 25.34$$

$$= \text{APPROXIMATELY } 25\ 3/8''$$

ANGLE OF MORTISE

ANGLE Ø (FROM TABLE) = 9 1/2°

TO FIND ANGLE Ø OF MORTISE

$$\text{TAN } Ø = \frac{AC}{CE} = \frac{4}{24} = 0.166$$

Fig. 2. The method of finding the length of a sawhorse leg and the angle, or inclination of side, of the mortise for the leg. To find φ by calculation, a table of natural trigonometric functions is necessary.

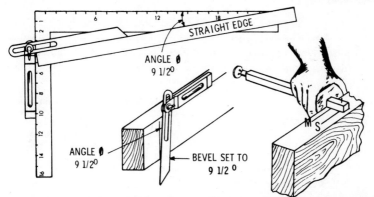

Fig. 3. The method of setting the bevel angle (ϕ) with the aid of a square and a straightedge. Place the straightedge on the square so that one side of the right triangle thus formed will be 24 inches (height of the sawhorse) and the other side will be 4 inches (distance to edge of leg from end of beam). Place the blade of the bevel against the straightedge, and place the stock against the side of the square; clamp the bevel to this angle, which is the proper slope for the side of the mortise.

CLAMPS

Frequently it is necessary to tightly press pieces of wood together that may have been mortised and tenoned, grooved and tongued, or simply glued. The bench vise is not always a convenient tool for this purpose, or its use may be required simultaneously for some other work. In such cases, clamps are used. There is quite a variety of clamps to meet the needs of the various kinds of work to be clamped. Clamps may be classed as:

1. Single-screw jaw.
2. Double-screw.
3. Beam.
4. Miter.
5. Chain.

The single-screw type consists of iron jaw clamps with a small to moderate opening, as shown in Fig. 4. Another type of single-screw clamp, usually called a bench bracket clamp, is shown in Fig. 5, with several different uses of this type of clamp. A form of double-screw clamp, sometimes called "hand screws," is shown

127

in Fig. 6. It is made entirely of wood and, if properly used, will be found satisfactory; however, a greenhorn or careless workman can easily destroy this type of clamp by abusing it, as illustrated in Fig. 6. Hand screws may be quickly opened or closed by

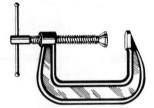

Fig. 4. A single-screw, iron-jaw clamp with a swivel head on the screw. Clamps of this type, often termed "C-clamps," are customarily used for clamping small pieces of wood and are made in openings that vary widely in size.

grasping the screw handles and revolving the clamp counterclockwise or clockwise, respectively (as viewed from the right hand).

For large openings, beam clamps are used (Fig. 7). Attached to one end of the beam is a head in which the screw works. A jaw that is arranged to slide on the beam is quickly adjusted and secured in any position. The chain type of clamp, as shown in

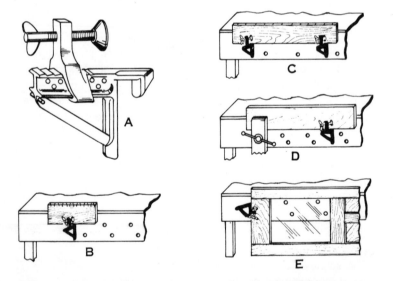

Fig. 5. A typical bench bracket clamp and its applications; A, the clamp; B, used for holding a short board so that is can be sawed at any angle; C, used for holding a long board—two brackets are used; D, used in conjunction with a bench vise; E, used for holding a door or window firmly in place while it is being worked.

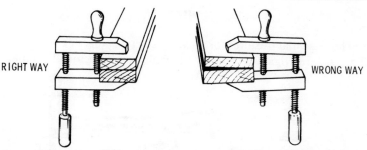

RIGHT WAY   WRONG WAY

Fig. 6. Right and wrong ways to use a hand screw. First, set the jaws to approximately the size of the material to be clamped. When placing the hand screw on the work, keep the points of the jaws slightly more open than the outer ends. Final adjustment of the inside screw will then bring the jaws exactly parallel, which is the proper position for clamping parallel work. Of course, if the work itself is wedge-shaped, the clamp jaws should conform so that equal pressure is applied at all points of contact. Since the screws are made of wood instead of iron or steel, proper judgment must be used with respect to the applied pressure.

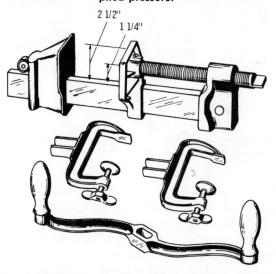

2 1/2"
1 1/4"

Fig. 7. A typical steel beam clamp. This type of clamp is designed for the heaviest class of hard wood work. The supports, as shown, are for fastening the clamp to a trestle or horse.

Fig. 8, is used for clamping built-up column work. The band clamp, Fig. 10, is a piece of nylon webbing attached to a ratchet. This type of clamp is used for clamping irregular shaped objects.

129

VISES

The essential features of a vise, such as the one shown in Fig. 9, are rigidity, weight, strength, and accurately fitting,

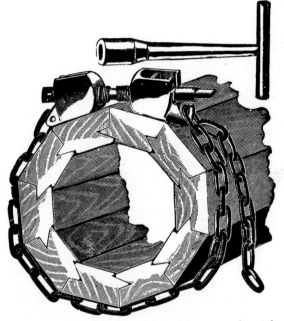

Fig. 8. A typical chain, or column, clamp. The 3/4-inch steel screw is threaded right and left, thus doubling the opening and closing speed.

smoothly working parts. Rigidity and weight are required to make effective the effort expended on the work held in the vise. The "anvil quality," or inertia, sufficient to effectively hold a piece of work solidly against the force of a blow is a most important qualification in a vise, and a suitable mass of iron or steel is just as necessary to supply this inertia as to supply strength against rupture. It is, of course, essential that a vise be strong enough to withstand any strain that may be legitimately put on it.

The types of vises usually fitted to the top of a workbench may be classed as:

1. Screw.

130

2. Quick-acting screw.
3. Parallel jaw.
4. Self-adjusting jaw.
5. Swivel bottom.

There is probably no tool in a shop subjected to more abuse than a vise. A fruitful cause of breakage is the clamping near one end of a long piece of work which may have a considerable overhang. Many times the operator, instead of hunting up a stick or other support to keep the free end from dropping, will attempt to hold it by excessive pressure between the vise jaws; if, in that condition, the operation involves any considerable hammering, the service exacted on that vise is most severe. One cause of minor breakage is the clamping of a hard piece of metal so that the pressure is concentrated on a small area near the margins or corners of the hardened jaw face; if the jaw is hardened enough to resist battering or indentation, a piece is almost sure to be broken out, leaving an unsightly notch. One common fault with

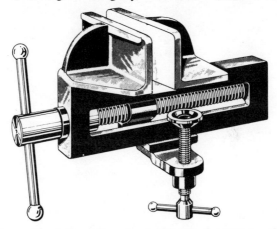

Fig. 9. The sectional view of a woodworker's vise that is suitable for medium-duty service.

vise users is the failure to keep the screw lubricated. The thread on many vise nuts has practically disappeared because of this. The front jaw should be occasionally detached from the vise, turned over, and the screw lubricated along its entire working length. When this is done at reasonable intervals, the screw and nut will last indefinitely. The use of vises having smooth faces for their gripping jaws is not nearly as extensive as it would be with a better comprehension of their capabilities.

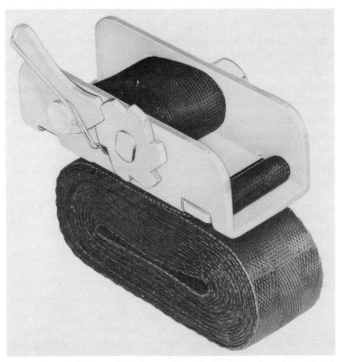

Fig. 10. A band clamp which is used to clamp irregular shaped objects.

SUMMARY

Holding tools are essential as part of workshop equipment. There are many tooling operations which require that the work be held rigid. As described in Chapter 5, the workbench is a part of the holding tool and should be anchored to the floor or wall in order to prevent movement.

Horses or trestles is a simple way to support work when large dimensions are being used. This type of labor-saving device is generally made from $2'' \times 4''$ or $2'' \times 6''$ stock for the cross beam, with a pair of $1'' \times 3''$ or $1'' \times 4''$ legs at each end.

An essential tool for any workbench is a vise. There are various types, such as screw, quick-acting screw, parallel jaw, self-adjusting jaw, and swivel bottom. The bench vise is not always a convenient tool to carry to a particular job, so in such cases clamps are used. Various types of clamps are available and may be classed as, single-screw, double-screw, beam, miter, and chain.

REVIEW QUESTIONS

1. What are some advantages in using a vise?
2. What is a sawhorse?
3. Why are clamps so important?
4. What is a chain clamp and when would it be used?
5. What is a hand screw?

CHAPTER 11

Toothed Cutting Tools

In almost any carpentry job, after the work has been laid out with guiding, marking, and measuring tools, and supported or held in position by a holding tool, the first cutting operation will, in most cases, be performed by a toothed cutting tool. The most important of this class of tools is the saw. Since sawing is hard work, the carpenter should not only know how to saw properly but should also know how to keep the saw in prime condition.

SAWS

There are many different kinds of saws, but the types of interest to the carpenter may be classed,

1. With respect to the type of cut, as:
   a. Crosscut.
   b. Ripsaw.
   c. Combined crosscut and ripsaw (interrupted tooth).
2. With respect to the shape of the blade, as:
   a. Straight back.
   b. Skew back.
   c. Thin back.
   d. Narrowed.
3. With respect to the reinforcement of the back, as:
   a. Half back.
   b. Full back.
4. With respect to service, as:
   a. Cabinet.
   b. Joiner.

    c. Miter.

    d. Stair.

    e. Floor.

    f. Buck or wood.

    g. Compass.

    h. Keyhole .

    i. Coping.

    j. Hack.

The common handsaw, as shown in Fig. 1, consists of a thin, flat blade of crucible steel having a row of teeth along one edge. A wooden handle is fastened to the large end by screws. A small size saw of this type is called a panel saw. The size of a saw

Fig. 1. A typical handsaw used in various woodcutting operations. The coarseness or fineness of a saw is determined by the number of teeth per inch. A coarser saw, properly set, is preferred for fast work on soft and green wood, whereas a finer saw is suitable for smooth, accurate cutting and for dry, seasoned wood. Ripsaws commonly have 5-1/2 to 6 teeth per inch, and crosscut saws have 7 or 8 teeth per inch.

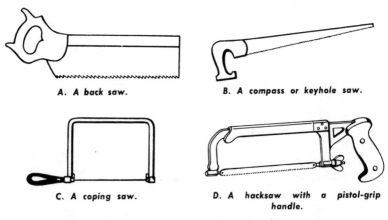

A. A back saw.

B. A compass or keyhole saw.

C. A coping saw.

D. A hacksaw with a pistol-grip handle.

Fig. 2. Four popular types of woodworking saws.

is determined by the length of the blade in inches. Handsaws range in size from 14 to 30 inches, as shown in Table 1.

The *back saw* (Fig. 2A) is a thin crosscut saw with fine teeth and is stiffened by a thick steel web through the entire length of the blade along the back edge. One popular size has a 12-inch blade length with 14 teeth per inch. It is used for making joints and in fine woodworking operations where great accuracy is desired.

Table 1. Saw Sizes

Size Inches	Panel						Hand	Rip	
	14	16	18	20	22	24	26	28	30

The *compass,* or *keyhole, saw* (Fig. 2B) has a small narrow blade with a pistol-grip handle and is commonly used for cutting along circular curves or lines in fine or small work.

The *coping saw* (Fig. 2C) is also used for cutting curves or circles in thin wood; it consists of a small narrow blade that is inserted in a sturdy metal frame in a manner similar to that used in the hacksaw.

The *hacksaw* (Fig. 2D), although used primarily for cutting metal, is a popular tool in any woodworking shop. There are two parts to a hacksaw—the frame and the blade. Common hacksaws may have either adjustable or solid frames, although the adjustable frame is generally preferred. Hacksaw blades of various types are inserted in these adjustable frames for different kinds of work; the blades vary in length from 8 to 16 inches. The blades are usually ½-inch wide and have from 14 to 32 teeth per inch. A hole at each end of the blade allows the blade to be hooked to the frame; a wing nut on one end regulates the blade tension, thus permitting various types of cutting actions.

Saw Teeth

The cutting edge of a handsaw is a series of little notches, all of the same size. On a crosscut saw, each side of the tooth is filed to a cutting edge like a little knife, as illustrated in Fig. 3. On a ripsaw, each tooth is filed straight across to a sharp square edge like a little chisel, as shown in Fig. 4.

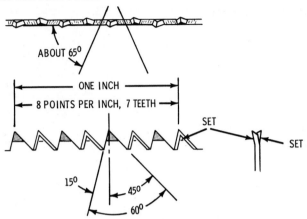

Fig. 3. *The side and tooth-edge views of a typical crosscut saw. This saw is used for cutting across the grain and has a different cutting action than that of the ripsaw. The crosscut saw cuts on both the forward and backward strokes.*

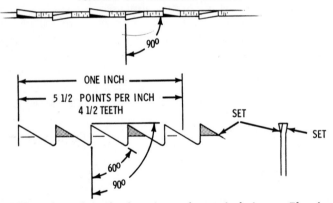

Fig. 4. *The side and tooth-edge views of a typical ripsaw. The ripsaw is used for cutting with the grain. Cutting is done only on the forward stroke.*

Set

The set of a saw is the distance that the teeth project beyond the surface of the blade. The teeth are "set" to prevent the saw from binding and the teeth from choking up with sawdust. In setting, the teeth are bent alternately, one to one side and the next to the other side, thus forming two parallel rows, or lines, along the edge.

Action of the Crosscut Saw

While each crosscut tooth resembles a little two-edge knife, it cuts quite differently. In early times, it was discovered that a knife blade must be free from nicks and notches to cut well. Then it could be pushed against a piece of wood, and a shaving could be whittled off. At about the same time, it was noticed that if the nicked knife were drawn back and forth across the wood, it would tear the fibers apart, making sawdust.

Fig. 5 shows the approved method of using the crosscut saw. As shown, an imaginary line through the saw, the arm, and the shoulder should be slightly to the left of the saw blade, thereby permitting

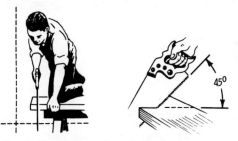

Fig. 5. The approved method of crosscutting.

a clear view of the cutting line and the stroke action. An angle of approximately 45° should be maintained between the crosscut saw and the face of the work.

The set of crosscut teeth makes them lie in two parallel rows. A needle will slide between them from one end of the saw to the other. When the saw is moved back and forth, the points, especially their forward edges, sever the fibers in two places, leaving a little triangular elevation that is crumbled off by friction as the saw passes through. New fibers are then attacked and the saw drops deeper into the cut.

Action of the Ripsaw

The teeth of the ripsaw are a series of little chisels set in two parallel rows that overlap each other. At each stroke, the sharp edge chisels off a little from the end of the wood fibers, as shown in Fig. 6. The teeth are made strong with an acute cutting angle, but the steel is softer than that of a chisel to enable the teeth to

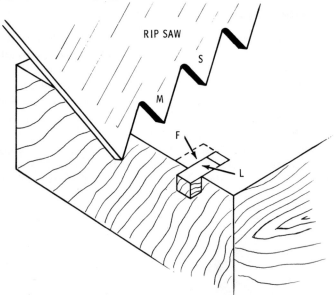

RIP SAW

S

M

F

L

Fig. 6. Action of the ripsaw. When the first tooth is thrust against the wood at an angle of approximately 45°, it chisels off and crowds out small particles of wood. Thus, tooth M will start the cut and take off piece L; tooth S will take off piece F, and so on.

be filed and set readily. Fig. 7 illustrates the proper position for cutting with a ripsaw.

Fig. 7. The proper position when sawing with the ripsaw.

Angles of Saw Teeth

The "face" of each crosscut tooth is slightly steeper than the back, thereby making an angle with the line of the teeth of approximately 66°. The compass teeth lean still further at an angle of 75°. The ripsaw face is at right angles (90°) to the line of the teeth. Its cutting edge is at right angles to the side of the blade. The angle of each tooth covers 60°. These angles are shown in Fig. 9.

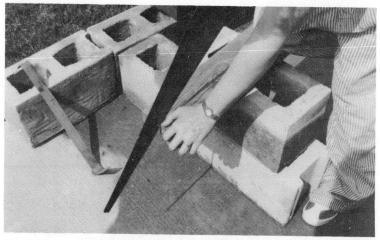

Fig. 8. To saw, grasp the wood with the left hand, and guide the saw with the thumb. Hold the saw lightly, and do not press it into the wood; simply move it back and forth, using long strokes.

FILES AND RASPS

By definition, a file is a steel instrument having its surface covered with sharp-edge furrows or teeth and is used for abrading or smoothing other substances, such as metal and wood. A rasp is a coarse file and differs from the ordinary file in that its teeth consist of projecting points instead of V-shaped projections extending across the face of the file.

Files are used for many purposes by woodworkers. Fig. 10 shown a variety of files. The taper file is adapted for sharpening hand, pruning, and buck saws. The teeth of the mill file leave a smooth surface. They are particularly adapted to filing and sharp-

TOOTHED CUTTING TOOLS

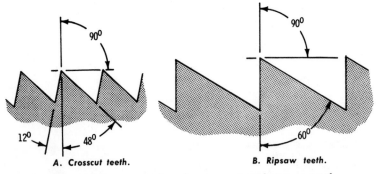

A. Crosscut teeth.   B. Ripsaw teeth.

Fig. 9. Angular proportions for crosscut and ripsaw teeth.

ening mill saws and mowing- and reaping-machine cutters. Rasps
are generally used for cutting away or smoothing wood or for
finishing off the rough edge left in a circular hole that has been
cut with the keyhole saw. The ordinary wood rasp is rougher or
coarser that that used by cabinetmakers. Wood files are usually
tempered to handle lead or soft brass.

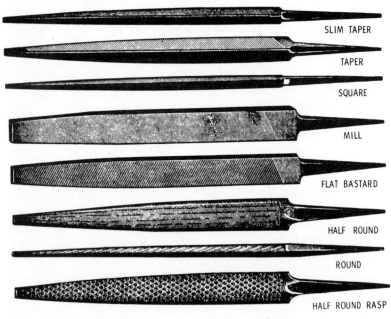

SLIM TAPER

TAPER

SQUARE

MILL

FLAT BASTARD

HALF ROUND

ROUND

HALF ROUND RASP

Fig. 10. Various types of files and rasps.

When drawing a file back between the cuts, do not allow it to drag, since it may be injured by about as much as when it is cutting. In using large rasps or files, whether for wood or metal, the work should be held in the vise or otherwise firmly fixed, because it is desirable to use both hands when possible. The handle of the tool should be grasped by one hand while the other hand is pressed, but not too heavily, on the end or near the end of the blade so as to lend weight to the tool and add to its powers of abrasion.

SANDPAPER

Sandpaper is a tough paper which is poured with glue and covered with sharp sand or some other abrasive material. It is used by the carpenter for the final finishing and smoothing of wooden surfaces. For use on sanding machines, such as planing-mill drums and discs and for floor-sanding machines, it is often furnished in ready-cut pieces. It is obtainable in 50-yard rolls in various widths from 6 to 48 inches; however, for shop use, it is often purchased in rolls with a 12-inch width or in 9"×11" sheets.

The papers used are a special manila and are strong and tough. The glues used were at one time the best grade of animal-hide glues, but at the present time, synthetic resins are used; many of these are water-resistant to some extent.

The cheapest abrasive is a natural flint, or a quartz, that is crushed and graded, but a more serviceable product is made from crushed natural garnet, which is quarried in some of the eastern states.

Many abrasive papers and cloths are covered with grains of manufactured abrasives, such as silicon carbide, boron carbide, and crystalline alumina. They are relatively expensive and are not especially well fitted for sanding wood, but they are used to a great degree in the metal trades and for some buffing operations. All abrasive papers and cloths are furnished in various grits which may or may not be numbered, or simply designated *extra fine* to *coarse*.

All species of woods do not respond equally well to machine sanding, and some of them are actually troublesome. The harder timbers, such as ash, oak, and hickory, sand evenly and smoothly, but some of the softer woods, such as gum, sycamore, yellow poplar, and some of the softer softwoods, may sand up to a fuzzy surface.

SUMMARY

Various types of saws are used in carpentry work. Since sawing is hard work, the carpenter should know how to saw properly and keep his saw in a sharp condition. There are many different kinds of saws, such as crosscut, rip, back, keyhole, coping, and hacksaw.

Files and rasps are tools used for smoothing other surfaces such as metal and wood. Files are also used to sharpen saws and other cutting tools. Various shapes of files and rasps are used in carpentry work, such as flat, taper, square, round, and half-round.

Sandpaper is a tough paper with abrasive material glued to one side. It is used for smoothing and final finishing of wood surfaces. Various sizes and grades are on the market, ranging from extra fine to coarse. Sandpaper generally is purchased in $9'' \times 11''$ sheets, but can also be obtained in rolls and discs for various sanding machines.

REVIEW QUESTIONS

1. How many teeth per inch in the average crosscut saw?
2. What is the difference between a crosscut saw and a ripsaw?
3. What is the difference between a coping saw and a keyhole saw?
4. Explain the purpose of a hacksaw.
5. Why must the saw teeth be set to project beyond the surface of the blade?

CHAPTER 12

Sharpening Saws

The term "sharpen" is used in its broad sense to include all the operations necessary to put a used saw into first-class condition. There are five steps in the sharpening of a saw; they are:

1. Jointing.
2. Shaping.
3. Setting.
4. Filing.
5. Dressing.

SHARPENING HANDSAWS

Handsaws are of two main types, namely, the crosscut and the ripsaw. The crosscut saw, as the name implies, is used to cut across the grain and to cut wet or soft woods. Ripsaws, on the other hand, are used to cut wood along the grain. They are both similar in construction, but ripsaws are slightly heavier; they also differ in the rake of the teeth. Other types of handsaws are: back saws, miter saws, dovetail saws, compass saws, keyhole saws, coping saws, etc.

When sharpening handsaws, the first step is to place the saw in a suitable clamp or saw vise, as illustrated in Fig. 1. In the absence of a good saw vise, a homemade clamp may easily be made in which the saw can be supported. The saw should be held tight in the clamp so that there is no noticeable vibration. The saw is then ready to be jointed.

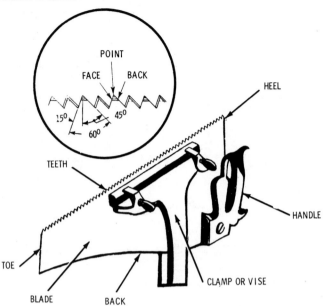

POINT

FACE   BACK

HEEL

15°   45°

60°

TEETH

HANDLE

TOE

CLAMP OR VISE

BLADE   BACK

Fig. 1. The method of fastening a handsaw in a saw clamp or vise.

Jointing

Jointing is done when the teeth are uneven or incorrectly shaped or when the teeth edges are not straight. If the teeth are irregular in size and shape, jointing must precede setting and filing. To joint a saw, place it in a clamp with the handle to the right. Lay a flat file lengthwise on the teeth, and pass it lightly back and forth over the length of the blade on top of the teeth until the file touches the top of every tooth. The teeth will then be of equal height, as shown in Fig. 2. Hold the file flat; do not allow it to tip to one side or the other. The jointing tool or handsaw jointer will aid in holding the file flat.

Shaping

Shaping consists of making the teeth uniform in width. This is normally done after the saw has been jointed. The teeth are filed with a regular handsaw file to the correct uniform size and shape. The gullets must be of equal depth. For the crosscut saw, the

146

TEETH TOO HIGH

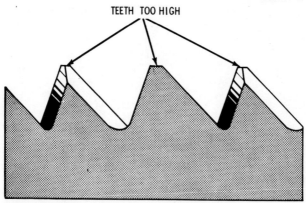

Fig. 2. The method of jointing saw teeth. Place the saw in a clamp with the handle to the right. Lay a mill file lengthwise flat on the teeth. Pass it lightly back and forth along the length of the teeth until the file touches the top of every tooth. If the teeth are extremely uneven, joint the highest teeth first, then shape the teeth that have been jointed and joint the teeth a second time. The teeth will then be the same height. Do not allow the file to tip to one side or the other; hold it flat.

front of the tooth should be filed at an angle of 15° from the vertical, while the back slope should be at an angle of 45° from the vertical, as illustrated in Fig. 3. When filing a ripsaw, the front of the teeth are filed at an angle of 8° with the vertical, and the back slope is filed at an angle of 52° with the vertical, as shown in Fig. 4. Some good workmen, however, prefer to file ripsaws with more of an angle than this, often with the front side of the teeth almost square, or 90°. This produces a faster-cutting saw, but, of

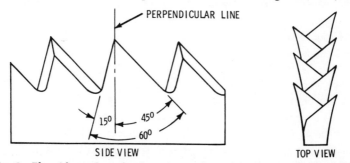

PERPENDICULAR LINE

15° 45°

60°

SIDE VIEW

TOP VIEW

Fig. 3. The side and tooth-edge views of a crosscut saw. The angle of a crosscut saw tooth is 60°, the same as that of a ripsaw. The angle on the front of the tooth is 15° from the perpendicular, while the back angle is 45°.

147

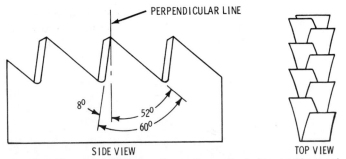

PERPENDICULAR LINE

SIDE VIEW

TOP VIEW

Fig. 4. The side and tooth-edge views of a typical ripsaw. The tooth of a ripsaw has an angle of 60°; that is, 8° from the perpendicular on the front and 52° on the back of the tooth.

course, it pushes harder, and it will grab when cutting at an angle with the grain.

When shaping teeth, disregard the bevel of the teeth, and file straight across at right angles to the blade with the file well down in the gullet. If the teeth are of unequal size, press the file against the teeth level with the largest flat tops until the center of the flat tops made by jointing is reached. Then move the file to the next gullet, and file until the rest of the flat top disappears and the tooth has been brought to a point. Do not bevel the teeth while shaping. The teeth, now shaped and of even height, are ready to be set.

Setting

After the teeth are made even and of uniform width, they must be set. Setting is a process by which the points of the teeth are bent outward by pressing with a tool known as a saw set. Setting is done only when the set is not sufficient for the saw to clear itself in the kerf. It is always necessary to set the saw after the teeth have been jointed and shaped. The teeth of a handsaw should be set before the final filing to avoid injury to the cutting edges. Whether the saw is fine or coarse, the depth of the set should not be more than one-half that of the teeth. If the set is made deeper than this, it is likely to spring, crimp, or crack the blade or break the teeth.

When setting teeth, particular care must be taken to see that the set is regular. It must be the same width along the entire length of the blade, as well as being the same width on both sides of the blade. The saw set should be placed on the saw so that the guides are positioned over the teeth with the anvil behind the

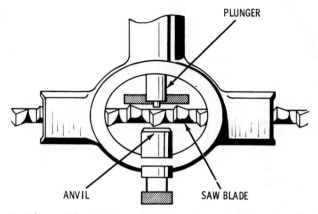

Fig. 5. The position of the saw set on the saw for setting the teeth.

tooth to be set, as shown in Fig. 5. The anvil should be correctly set in the frame, and the handles should be pressed together. This step causes the plunger to press the tooth against the anvil and bend it to the angle of the anvil bevel. Each tooth is set individually in this manner.

Filing

Filing a saw consists of simply sharpening the cutting edges. Place the saw in a filing clamp with the handle to the left. The bottom of the gullets should not be more than ½ inch above the jaws of the clamp. If more of the blade projects, the file will chatter or screech. This dulls the file quickly. If the teeth of the saw have been shaped, pass a file over the teeth, as described in jointing, to form a small flat top. This acts as a guide for the file; it also evens the teeth.

To file a crosscutting handsaw, stand at the first position shown in Fig. 6. Begin at the point of the saw with the first tooth that is set toward you. Place the file in the gullet to the left of this tooth, and hold the handle in the right hand with the thumb and three fingers on the handle and the forefinger on

149

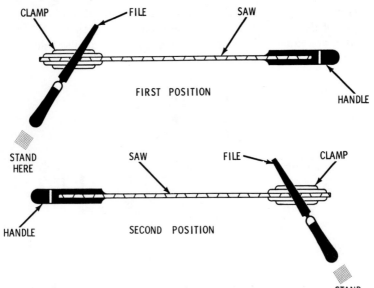

CLAMP FILE SAW

FIRST POSITION

HANDLE

STAND HERE

SAW FILE CLAMP

HANDLE SECOND POSITION

STAND HERE

Fig. 6. Standing positions for filing a crosscut saw. The saw clamp should be moved along the blade as filing progresses.

top of the file or handle. Hold the other end of the file with the left hand, the thumb on top and the forefinger underneath. The file may be held in the file-holder guide, as shown in Fig. 7.

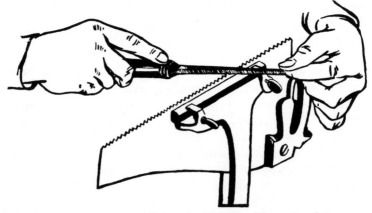

Fig. 7. The method of holding the file when filing a handsaw.

150

The guide holds the file at a fixed angle throughout the filing process while each tooth is sharpened.

Hold the file directly across the blade. Then swing the file left to the desired angle. The correct angle is approximately 65°, as shown in Fig. 8. Tilt the file so that the breast (the front side of the tooth) side of the tooth may be filed at an angle of approximately 15° with the vertical, as illustrated in Fig. 3. Keep the file level and at this angle; do not allow it to tip upward or downward. The file should cut on the push stroke and should be raised out of the gullet on the reverse stroke. It cuts the teeth on the right and left on the forward stroke.

File the teeth until half of the flat top is removed. Then lift the file, skip the next gullet to the right, and place the file in the second gullet toward the handle. If the flat top on one tooth is larger than the other, press the file harder against the larger

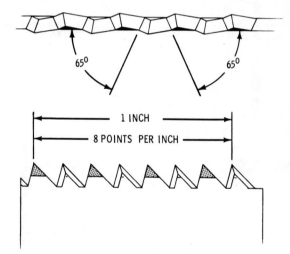

Fig. 8. *The side angle at which to hold the file when filing a crosscut saw having eight points per inch.*

tooth so as to cut that tooth faster. Repeat the filing operation on the two teeth which the file now touches, always being careful to keep the file at the same angle. Continue in this manner, placing the file on every second gullet until the handle end of the saw is reached.

151

Turn the saw around in the clamp with the handle to the left. Stand in the second position, and place the file to the right of the first tooth set toward you, as shown in Fig. 6. This is the first gullet that was skipped when filing from the other side. Turn the file handle to the right until the proper angle is obtained, and file away the remaining half of the flat top on the tooth. The teeth that the file touches are now sharp. Continue the operation until the handle end of the saw is reached.

When filing a ripsaw, one change is made in the preceding operation; the teeth are filed straight across the saw at right angles to the blade. The file should be placed on the gullet so as to file the breast of the tooth at an angle of 8° with the vertical, as shown in Fig. 4. Stand in the positions shown in Fig. 9. When sharpening a ripsaw, file every other tooth from one side. Com-

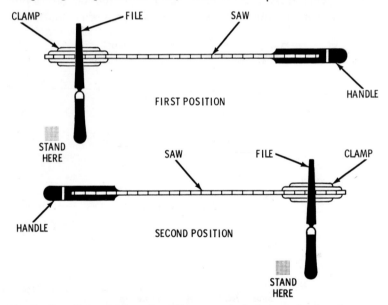

CLAMP   FILE   SAW

HANDLE

FIRST POSITION

STAND
HERE

SAW   FILE   CLAMP

HANDLE

SECOND POSITION

STAND
HERE

Fig. 9. Standing positions for filing a typical ripsaw. Again, the saw clamp must be moved along the blade as filing progresses.

pare Fig. 10 with Fig. 8 to see the difference in the teeth angles. Then turn the saw around, and sharpen the remaining teeth as described in the preceding paragraphs. When filing teeth, care must be taken in the final sharpening process to file all the teeth to the same size and height, otherwise the saw will not cut satisfactorily. Many good saw filers file ripsaws from only one side, taking care that the file is held perfectly horizontal. For the beginner, however, turning the saw is probably the most satisfactory method.

Dressing

Dressing of a saw is necessary only when there are burrs on the side of the teeth. These burrs cause the saw to work in a ragged fashion. They are removed by laying the saw on a flat surface and running an oilstone or flat file lightly over the side of the teeth.

SUMMARY

Success in woodworking depends on proper tool maintenance, which includes the tool sharpening process. There are five steps which should be considered when sharpening saws; they are jointing, shaping, setting, filing, and dressing. Filing a saw consists of simply sharpening the cutting edges.

Jointing is done when the teeth on the blade are uneven or incorrectly shaped or when the teeth edges are not straight. The high spots or high teeth are filed down and then sharpened. Shaping a saw consists of making the teeth uniform. The teeth are filed to the correct size and shape. The gullets must be of equal depth.

After the teeth are made even and uniform in width, they must be set. This is accomplished with the use of a tool known as a saw set. It is necessary to set a saw when the teeth have been jointed and shaped. The saw must be placed in the clamp properly and held solid.

REVIEW QUESTIONS

1. What are the five steps used in sharpening a saw?
2. What is the heel on a hand saw?
3. What is jointing a saw?
4. How is jointing and shaping accomplished?
5. Explain the process of dressing a saw.

CHAPTER 13

Sharp-Edged Cutting Tools

One reason for so many poor results in carpentry is the use of dull cutting tools. Not only should the edge be whetted or honed as soon as any sign of dullness is observed, but the tools should always be kept perfectly clean and free from dust. The tools to be discussed in this chapter are considered as hand-guided sharp-edged cutting tools, such as chisels, draw knives, etc., as distinguished from striking tools (hatchets) and self-guided tools (planes).

CHISELS

In carpentry, the chisel is an indispensable tool. It is, however, one of the most abused tools, because it is often used for prying open cases and even as a screwdriver, although it is adapted solely for cutting wood surfaces. A chisel consists of a flat, thick piece of steel with one end ground at an acute bevel to form a cutting edge and the other end provided with a handle, as shown in Fig. 1.

Chisels may be classed:
1. With respect to duty or service, as:
   a. **Paring.**
   b. Firmer.
   c. Framing or mortise.
   d. Gouge.
2. With respect to the length of the blade, as:
   a. Butt.
   b. Pocket.
   c. Mill.
3. With respect to the abnormal width of the blade, as slick.
4. With respect to the edges of the blade, as:

    a. Plain.

    b. Bevel.

5. With respect to the method of attaching the handle, as:

    a. Tang.

    b. Socket.

6. With respect to the shape of the blade, as:

    a. Flat.

    b. Round (gouge).

    c. L (corner).

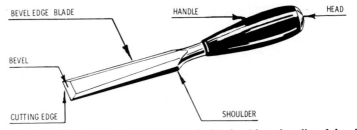

Fig. 1. A typical general-purpose wood chisel with a handle of hard plastic that possesses a tremendous resistance to breakage.

Paring Chisel

This is a light-duty tool for shaping and preparing relatively long planed surfaces, especially in the direction of the grain of the wood. The paring chisel (Fig. 2) is manipulated by a steady sustained pressure of the hand; it should not be driven by the blows of a hammer or other similar tool.

Firmer Chisel

The term "firmer" implies a more substantial tool than the paring chisel, and is adapted to medium-duty work. The firmer chisel (Fig. 2) is a tool for general work and may be used either for paring or light mortising; it is driven by hand pressure in paring and by blows from a mallet in mortising.

NOTE: A hammer or other tool should not be used to drive a chisel—use only a mallet. Wood to wood in driving is the only

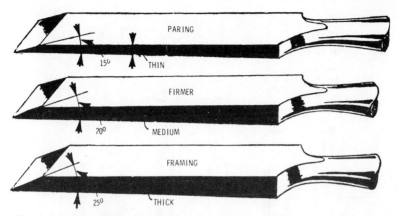

Fig. 2. *Various chisels classified with respect to duty. They are grouped as paring for light duty, firmer for medium duty, and framing for heavy duty.*

satisfactory method for driving chisels. An exception to this rule is the framing chisel, which has a special handle.

Framing or Mortise Chisel

This is a heavy-duty tool that is adapted to withstand the severe strain required in framing where deep cuts are necessary. In the best construction, an iron ring is fitted to the end of the handle to protect it from splitting, thus permitting the use of a heavy mallet when driving the tool into the wood. The framing chisel (Fig. 2) may be used as a deck chisel in ship and boat construction.

Slick

Chisels with wide blades and long handles are called slicks. They may have widths of 2½, 3, 3½, and 4 inches. Slicks are used principally in heavy framing work, and as the name implies, to smooth and true tenons and other broad joints. They are driven with the hands only and should not be driven with a mallet or hammer. The preliminary work is usually done with an adze.

Gouge

This is a chisel with a hollow-shaped blade for scooping or cutting round holes. There are two kinds of gouge chisels, the out-

157

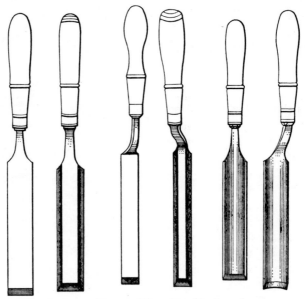

Fig. 3. Several types of framing chisels and gouges.

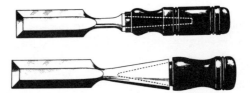

Fig. 4. Typical tang and socket chisels, respectively. The terms "tang" and "socket" are derived from the fact that the shank of the tang chisel has a point which is fitted into the handle. This point is called a tang, hence the name "tang chisel." In the socket chisel, the shank of the chisel is made like a cup, or socket, with a handle fitted into it; thus the chisel is called a "socket chisel."

side bevel and the inside bevel, as shown in Fig. 3. The outside bevel is the more common of the two.

Corner Chisel

This type of chisel has an L-shaped blade and is used in clearing out corners and angles (90° and over), in squaring holes, and in general for a V-cut required in pulley stiles or in handrail mouldings.

Tang and Socket Chisels

According to the method by which the blade and handle are joined, chisels are called tang or socket. The difference in the two types is shown in Fig. 4. The tang chisel has a projecting part, or tang, on the end of the blade which is inserted into a hole in the handle. The reverse method is employed in the socket chisel; that is, the end of the handle is inserted into a socket on the end of the blade. The term "socket firmer," as applied to a firmer chisel having a socket end, does not mean (as generally supposed by some carpenters) "hit it firmer," although that is what actually happens in operation; the blows tend to drive the handle firmer into the tapered socket.

Butt, Pocket, and Mill Chisels

This classification relates simply to the relative lengths of the blades. The regular lengths, shown in Fig. 5, are approximately as follows: butt, 2½ to 3¼ inches, pocket 4 to 5 inches, and mill, 8 to 10 inches.

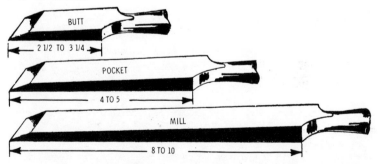

Fig. 5. Wood chisels classified with respect to the length of the blade. The blade width may vary, depending on the type of work to be performed. The lengths given in the illustration are in inches.

How to Select Chisels

A chisel should be absolutely flat on the back (the side should not be beveled). An inferior chisel is ground off on the back near the cutting edge, with the result that, in use, it tends to follow the grain of the wood, splitting it off unevenly, since the user cannot properly control his tool. The flat back allows the chisel to take off the finest shaving, and where a thick cut is desired, it will not strike too deep. This is an important point to be looked for in good chisels.

The best chisels are made of selected steel with the blade widening slightly toward the cutting edge. The blades are oil-tempered and carefully tested. The ferrule and blade of the socket chisel are so carefully welded together that they practically form a single piece. Socket chisels are preferred to the tang type by most carpenters, because they are stronger and the handles are less apt to split. Beveled edges are preferable to plain blades, because they tend to drive the tool forward and also have greater clearance.

The chisels commonly carried on construction jobs these days are quite short—only 3- or 3½-inch beveled blades and metal-capped handles of a tough plastic that will take an amazing amount

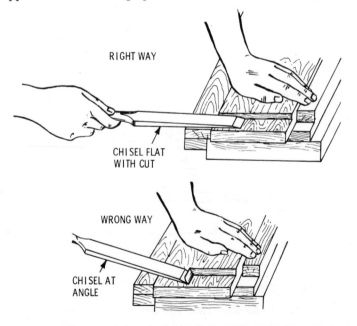

RIGHT WAY

CHISEL FLAT
WITH CUT

WRONG WAY

CHISEL AT
ANGLE

Fig. 6. The right and wrong ways to use a paring chisel. The flat side of the chisel should face the cut and be parallel with it.

of abuse. The handles are cast and moulded in place and are intended to be driven with a hammer. Such chisels are too short for a good job of paring or deep mortising but are serviceable for general heavy-duty use and for door-butt and similar mortising.

One excellent handle for socket-firmer chisels is sometimes, but rarely, available. It is built-up of sole-leather rings around a steel rod core. These handles are almost indestructible and may be driven with a hammer.

The butt chisel, because of its short blade, is adapted for close accurate work where not much power is required. It is particularly suited for putting on small hardware, which does not necessarily require the use of a hammer. The butt chisel may be used almost like a jackknife with the hand placed well down on the blade toward the cutting edge. The short blade and handle make it convenient for carrying in the pocket. Chisels are usually ground sharp and hand honed and are ready for use when sold.

How to Care For and Use Chisels

In order to do satisfactory work with chisels, the following instructions should be carefully noted and followed:

1. Do not drive the chisel too deep into the work; this requires extra pressure to throw the chips.
2. Do not use a firmer chisel for mortising heavy timber.
3. Keep the tool bright and sharp at all times.
4. Protect the cutting edge when not in use.
5. Never use a chisel to open boxes, to cut metal, or as a screwdriver or putty knife.

A skilled mechanic does not need the advice given here. It is intended for those users of tools who learn through experience that the unsatisfactory results frequently obtained are due to the fact that the user and not the tool is defective. Take particular note of Figs. 6 and 7, which illustrate the use of the paring, the firmer, and the framing chisels. Fig. 8 also shows the proper use of chisels.

How to Sharpen Chisels

When honing a chisel, use a good grade of oilstone. Pour a few drops of machine oil on the stone, or, if you have no machine oil,

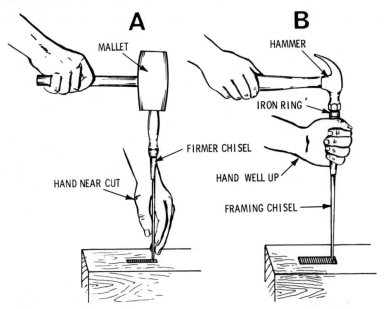

Fig. 7. Mortising with a chisel; A, using a firmer chisel for light mortise work (note the position of the hand and the type of driving tool used); B, using a framing chisel for cutting a large mortise (note the position of the hand and the type of driving tool used).

use lard or sperm oil. The best results are obtained by using a carborundum stone. The carborundum cuts faster than most other abrasives, but the edge will not be as smooth and keen as when a natural oilstone is used.

Hold the chisel in the right hand, and grasp the edges of the stone with the fingers of the left hand to keep it from slipping. A better method is to place the stone on a bench and block it so it cannot move; both hands will then be free to use for honing. In this case, grasp the chisel in the right hand where the shoulder joins the socket. Place the middle and forefinger on the blade near the cutting edge. Rub the chisel on the oilstone away from you; be careful to keep the original bevel.

Never sharpen the chisel on the back or flat side; this should be kept perfectly flat. For paring, the taper should be long and thin— approximately 15°, as shown in Fig. 2. The longer the bevel on the cutting edge, the easier the chisel will work, and the easier it

A. The use of a firmer chisel for light mortise work.   **B. The use of a paring chisel.**

C. The method of cutting a concave curved corner.

Fig. 8. Using chisels for various types of work.

is to hone. A firmer chisel should be ground at an angle of not less than 20°; an angle of 25° is recommended for a framing chisel. When honing a chisel, the taper should be carefully maintained, and, unless the back is kept flat, it will be impossible to work to a straight line. Bevel-edge chisels are more easily sharpened than the plain-edge type, because there is not as much steel to be removed.

If the chisel is badly nicked, it will have to be ground before honing. Not many quality chisels can be filed. Do not overheat or

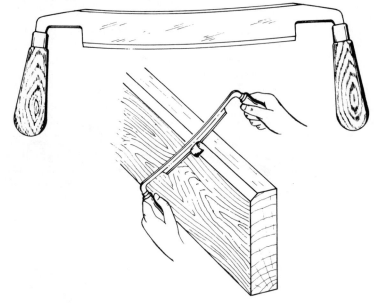

Fig. 9. A typical drawknife and its method of use.

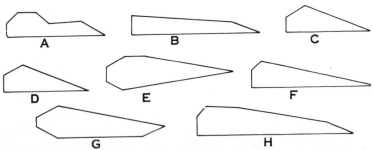

Fig. 10. Cross sections of various drawknife blades; A. Carpenters' razor blade; B. Carpenters' light blade; C. Carriage makers' narrow blade; D. Coach makers' razor blade; E. Wagon makers' heavy blade; F. Shingle shave, heavy blade; G. Saddle tree shave, heavy blade; H. Spar shave, heavy blade.

injure the temper of the chisel, and be sure to keep the original taper of the bevel. After grinding, hone the chisel on an oilstone as instructed.

DRAWKNIFE

This tool consists of a large sharp-edged blade with a handle at each end, usually at right angles to the blade, as shown in Fig. 9. It is used for trimming wood by drawing the blade toward the user. When the blade is sharp, and some degree of force is applied, it does its work quickly and efficiently.

The tool was formerly used to a great extent for the rapid reduction of stock to an approximate size, an operation that is now to a greater extent performed by sawing or planing machines. The drawknife is, however, quite effective on narrow surfaces that must be considerably reduced; the work is first seared with an adze or hand axe. Drawknives are made with cross sections of various shapes, thereby adapting them for a variety of service, as illustrated in Fig. 10.

SUMMARY

One of the main reasons for poor results in woodworking is the use of dull cutting tools; that is, neglect in keeping tools sharp. Not only should the cutting edge be whetted or honed as soon as any sign of dullness is observed, but the tools should always be kept perfectly clean and free from rust. Sharp-edged cutting tools may be divided into several classes, such as chisels, planes, drawknives, hatchets, and axes.

When honing a chisel, use a good grade of oilstone. Pour a few drops of machine oil on the stone and rub the chisel away from you, being careful to keep the original bevel. Never sharpen a chisel on the back or flat side—this should be kept perfectly flat. The best chisels are made of selected steel with the blade widening slightly toward the cutting edge.

REVIEW QUESTIONS

1. What is the difference between a tang chisel and a socket chisel?
2. What is a paring chisel?
3. What is a drawknife?
4. What are the three basic blade lengths for chisels?
5. What is a corner chisel?

CHAPTER 14

Rough Facing Tools

The tools described in this section are classed by some carpenters as striking tools, because the work is done with a series of blows. They might also be classed as inertia tools, because when the rapidly moving tool strikes the wood, its inertia, or property of matter that causes it to resist a change from motion to rest, drives the tool into the wood. The cut produced by this method is rough as compared with that made by other tools; hence the classification "rough facing tools."

HATCHET

This is a general utility tool that is familiar to all. In framing timber, it can be used as a hammer, or for sharpening stakes, cutting timber to a rough size, or splitting wood. Various types of hatchets are shown in Fig. 1.

BROAD HATCHET OR HAND AXE

This is simply a large hatchet with a broad cutting edge. Ordinarily it is grasped with the right hand at a distance of approximately one-third from the end of the handle, but the position of the hand will be regulated to a great measure by the material to be cut, that is, by the intensity of the blow, as illustrated in Fig. 3. Thus, to deliver a heavy blow, the handle is grasped close to the end, and for a light blow, it is held nearer to the head of the axe.

ROUGH FACING TOOLS

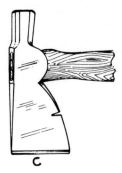

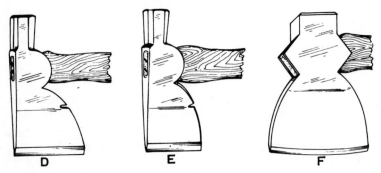

Fig. 1. Various types of hatchets. A. Shingling; B. Claw; C. Barrel; D. Half; E. Lath; F. Broad (hand axe).

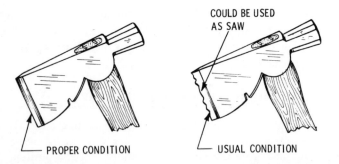

COULD BE USED
AS SAW

PROPER CONDITION

USUAL CONDITION

Fig. 2. The proper and usual conditions of a hatchet or axe cutting edge. There should not be a single nick on the cutting edge; unless the edge is in the proper condition, the tool is useless for any purpose except to cut firewood.

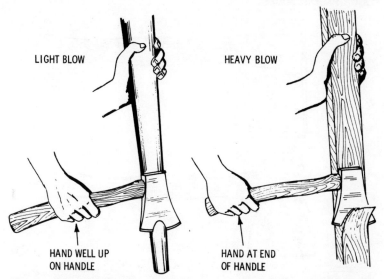

LIGHT BLOW

HEAVY BLOW

HAND WELL UP
ON HANDLE

HAND AT END
OF HANDLE

Fig. 3. Grasp the handle of the hand axe approximately halfway between the ends to strike a light blow and at the end of the handle to obtain the necessary swing for a heavy blow.

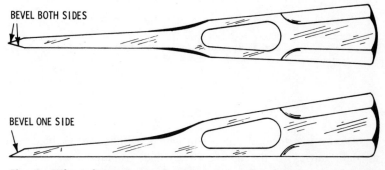

BEVEL BOTH SIDES

BEVEL ONE SIDE

Fig. 4. When sharpening a hatchet or axe, bevel both sides of the cutting edge for general use; for hewing a line, bevel only one side.

AXE

This tool is similar to the hand axe but is of a larger size and a longer handle and is intended for heavy cutting using both hands. A typical axe is shown in Fig. 5 at the bottom of the illustration.

169

Fig. 5. Typical hatchets and an axe.

ADZE

An adze is a form of hatchet in which the edge of the blade is at right angles to the handle. The blade is curved or arched toward the end of the handle, thus permitting an advantageous stroke of the tool while the operator is standing over the work. The edge is beveled on the inside only, and the handle can be removed when it is necessary to grind the tool.

When sharpening an adze, the tool should be traversed across the face of the stone; hold it at the proper bevel angle until all the nicks have been taken out. Then to secure a keen edge, rub it with a slipstone. It is important when sharpening an adze to bevel only on the inside.

SUMMARY

A hatchet can be used as a hammer, for sharpening stakes, cutting timber to a rough size, or splitting wood. The hatchet is also used in roof shingling.

There are various types of hatchets, such as shingling, claw, barrel, half, and lath. The cutting edge must, at all times, be free of nicks in order to be a useful tool.

REVIEW QUESTIONS

1. Name the various types of hatchets.
2. What is an adze?
3. In what type of work would a hatchet be used?
4. What is the difference betweeen an axe and a hatchet?
5. How are hatchets used in framing work?

171

CHAPTER 15

Smooth Facing Tools

The tools under this classification are those sharp-edged cutting tools in which the cutting edge is guided by the contact of the tool with the work instead of being guided by hand. For example, consider a plane as distinguished from a chisel. The plane, being positively guided, gives a smooth cut in contrast with the rough cut obtained by the hand-guided chisel—hence the term "smooth facing tools." These tools are essentially chisels set in appropriate frames, so that the contact of the frame with the work during the movement of the tool will give a positive guide to the cutting edge, thus resulting in a smooth cut.

THE SPOKESHAVE

This tool resembles a modified drawknife, whose blade is set in a boxlike frame which forms a positive guide. The blade is adjustable like a plane to govern the thickness of the cut. Spokeshaves may be made of wood or metal, and they have obtained their name from the fact that they were once used in the making of wagon spokes. They may be used to smooth curved edges and to round irregular surfaces, as shown in Fig. 1. Spokeshaves are made with cutters of various shapes—straight, hollow, round, angular, etc. A typical spokeshave is shown in Fig. 2.

The flat-bottom spokeshave is used on convex and concave surfaces where the curves have a long sweep; it is also used to chamfer or round edges. The hollow-bottom, or concave-bottom, spokeshave is used for rounding edges that have small convex sweeps, and the convex-bottom spokeshave is employed to cut concave curved edges that have small sweeps.

Fig. 1. The method of using the spokeshave. The spokeshave cuts when pushed away from the user, as indicated by the arrow. Be careful to work in the direction in which the tool cuts without tearing the grain. The spokeshave is also used to chamfer and cut edges.

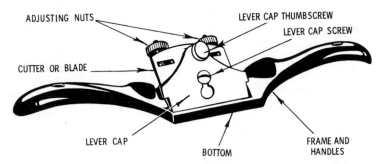

Fig. 2. A typical spokeshave. This is a light-weight, handy tool for use on concave, curved edges that have large or small sweeps. The wing-nut adjustment is on the cutter cap.

Spokeshave cutters may be sharpened by removing the blade from the stock and rubbing it on the inside with a flat slip of oilstone; lightly rub the outside of the blade on an ordinary oilstone. To more firmly hold the small blade, place it into a saw kerf made across the end of a small, flat piece of wood, with the edge of the blade projecting beyond the wood. The piece of wood should be beveled to allow the blade to lie on the stone at the proper angle. It may them be sharpened like a plane iron.

PLANES

A plane is a tool for smoothing boards or other surfaces of wood. It consists of a stock (usually made of wood or iron or a combination of both), from the underside, or face, of which projects slightly the steel cutting edge. The cutting edge of a plane, which inclines backward, is called the plane iron. An aperture in the front provides for the escape of the shavings which are produced when planing.

The plane is essentially a finishing tool, and, while it is adapted for use in bringing wood surfaces to the desired thickness, it will produce this result only gradually as compared to a chisel or hatchet. For this reason, it is normally the last tool to be used in finishing a wood surface.

There are a multiplicity of planes (some of which are shown in Fig. 3) to meet the varied requirements, such as:

1. Jack plane.
2. Fore plane.
3. Jointer plane.
4. Smoothing plane.
5. Block plane.
6. Molding plane.
7. Rabbet plane.
8. Fillister plane.
9. Grooving plane.
10. Router plane.
11. Chamfer plane.

Jack Plane

This plane is intended for heavy rough work. It is generally the first plane used in preparing the wood; its purpose is to remove irregularities left by the saw and to produce a fairly smooth surface. The jack plane is long and heavy enough to make it a powerful tool, so that it will remove a considerable chip with each cut. The cutting edge of the plane iron is ground slightly rounded; this form is best adapted for roughing. If properly sharpened, the jack plane may be used as a smoothing plane, or as a jointer on small work, because it is capable of doing just as good work.

Fore Plane

This plane is designed for the same purpose as the jointer plane, that is, to straighten and smooth the rather rough and irregular

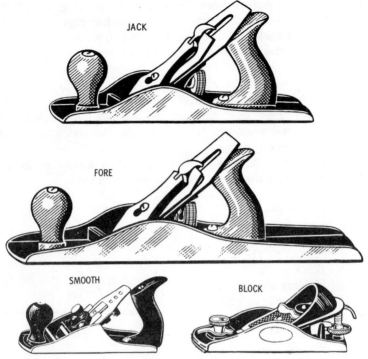

Fig. 3. Various types of planes used by woodworkers.

cut of the jack plane. Since the fore plane is shorter than the jointer (usually 18 inches in length), it is easier to handle, especially for a journeyman carpenter, and it may also be used as a jack plane. If a carpenter does not have both a jack plane and a jointer, he can make a fore plane serve for both, although it will not give as good service as either of the other two in the work for which they are adapted. The plane iron of the fore plane is sharpened to a straight line and is set for a finer cut than that of the jack plane.

Jointer Plane

The great length and weight of these planes keep the cutter from tearing the wood, and, with the cutter set for a fine cut, it is the plane to use for obtaining the smoothest finishes. These planes will true up better than other types of planes.

In this country, the word "jointer" is applied to planes that range in size from 22 to 30 inches. The length of the plane determines the straightness of the cut. Thus, a smoothing plane, because of its short length, will follow the irregularities of an uneven surface, taking its shavings without interruption, whereas, a fore or jointer plane similarly applied will first touch only the high spots, progressively lengthening the cuts until, on reaching the lowest spots, a continuous shaving will be taken. The final cut will approach a true plane surface in degree depending on the length of the tool and the length of the irregularities or undulations of the surface originally present. The cutting edge of a jointer plane is ground straight and is set for a fine cut.

Smoothing Plane

The small length of this plane (usually about 8 inches) adapts it for finishing uneven surfaces; because of its small size, it will find its way into minor depressions of the wood without taking off much material. In this respect, the smoothing plane differs from

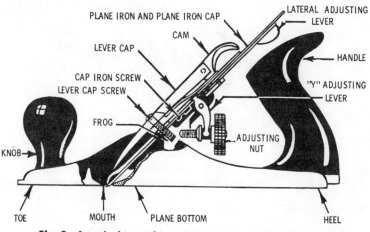

Fig. 5. A typical smoothing plane and its component parts.

the jointer plane; although both are finishing planes, the jointer plane is used for finer work. A typical smoothing plane is shown in Fig. 4.

177

Block Plane

This type of plane, shown in Fig. 5, is the smallest plane made (length 4 to 7 inches). It was designed to meet the demand for a plane that may easily be held in one hand while planing across the grain. The block plane is used almost exclusively for planing across the grain; therefore, no cap iron is necessary to break the shavings, since they are only chips.

The bevel of the plane iron is turned up instead of down. Because of its size, the block plane is usually operated with one hand, with the work held by the other hand. Therefore, as distinguished from this method of using, other planes are called bench planes. The angle of the plane iron for block planes is much smaller than that for bench planes. This angle is 20° for soft woods and 12° for hard woods; planes with the iron set at 12° are called "low angle" block planes.

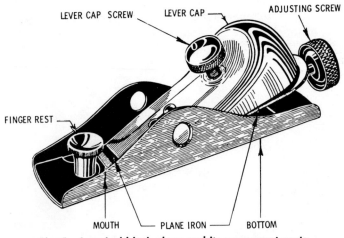

LEVER CAP SCREW   LEVER CAP   ADJUSTING SCREW

FINGER REST

MOUTH   PLANE IRON   BOTTOM

Fig. 5. A typical block plane and its component parts.

Rabbet Plane

In this type of plane, as shown in Fig. 6, the plane iron projects slightly from the side as well as from the bottom of the plane. There are various forms of rabbet planes available, each suitable for different types of cuts. With a tool of this type, the edge of a board can be cut so as to leave a rabbet or "sinking" (like a step) along its length to fit over and into a similar indentation cut in the

178

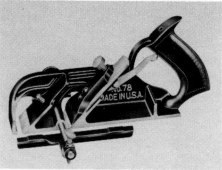

Fig. 6. A typical rabbet plane.

edge of another board. Rabbet planes are adapted to cut with or across the grain according to the setting of the iron.

Fillister Plane

Because of the difficulty of using a rabbet plane with accuracy, it is usual to use a fillister plane, which is a modified rabbet plane with an adjustable fence or guide that may be attached to the side of the plane and may be regulated according to the width of the rabbet to be made. A screw stop is also placed on the side of the plane farthest from the operator to regulate the depth of the cut.

The word "fillister" is defined in carpentry as a kind of moulding plane designed for making grooves or forming rabbets, as in window sashes; the rabbet is on the sash bar to receive the edge of the glass and the putty.

Grooving Plane

This plane (sometimes called a trenching plane) is used for cutting grooves across the grain. It has a rabbeted sole; the cutters are in the tongue portion, which is usually ½ inch deep and varies from ¼ to 1⅛ inches. A screw stop adjusts the depth of the cut, and a double-toothed cutter separates the fibers in front of the iron. A typical grooving plane is shown in Fig. 7.

Router

Planes of this type are employed for surfacing the bottom of grooves or other depressions parallel with the general surface of

179

the wood. The closed-throat type is the ordinary form of router; the open-throat type is an improved design, giving more freedom

Fig. 7. A typical grooving, or plow, plane, especially designed for weatherstrip grooving.

for chips and a better view of the work and cutter. Both types are shown in Fig. 8. The open-throat router has an attachment for regulating the thickness of the chip and a second attachment for closing the throat for use on narrow surfaces. The bottoms of both

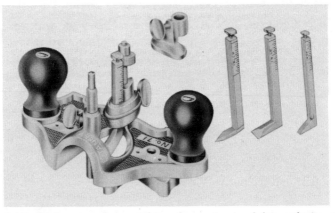

Fig. 8. Typical router planes. This plane is used for surfacing the bottom of grooves or other depressions parallel to the work.

styles are designed so that an extra wooden bottom of any size desired can be screwed on, thereby enabling the user to rout large openings.

Chamfer Plane

This plane produces a regular chamfer on the salient angle of a board. An adjustable stop sliding in the mouth of the plane regulates the width of the chamfer, which is limited to an angle of 45° with the sides. An ordinary plane can be arranged to cut a

180

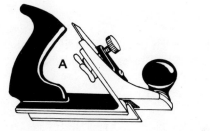

Fig. 9. An adjustable chamfer plane for chamfer or stop-chamfer work. The front piece can be set for different chamfer sizes. Front A may be detached and front B may be substituted, thereby permitting the plane to be worked closer into corners.

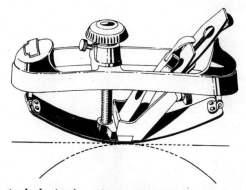

Fig. 10. A typical circular plane. The flexible steel face may be adjusted to the required arc, either concave or convex, for planing curved surfaces; it is accurately set and firmly held in place by the knob and setscrew.

chamfer by fastening strips of hardwood to its sole. A chamfer at any angle can be cut by providing the strips with the desired angle.

Universal Plane

A universal, or combination, plane is employed to produce a great number of mouldings. This type of plane is illustrated in Fig. 11. Variously formed cutters may be inserted or exchanged as conditions require. It may also be used for tongue-and-groove work, slitting, and sash work. Fig. 12 shows the variety of cutters used in conjunction with the universal plane.

Plane Irons or Cutters

The so-called plane iron which does the cutting is similar to a chisel but differs in that its sides are parallel and the thickness is less than that of the chisel blade. Plane irons are classed:

1. With respect to thickness, as:
   a. Heavy.
   b. Thin.
2. With respect to the shape of the cutting edge, as:
   a. Curved.
   b. Straight (square).
   c. Skew (oblique).
   d. Toothed.
3. With respect to provisions for breaking the chips, as:
   a. Single (Fig. 13).
   b. Double (Fig. 14).

Heavy plane irons are usually No. 12 gauge, whereas the medium or thin plane irons are usually No. 14 gauge. The heavy plane iron offsets the tendency found in spring-cap planes to vibrate; the additional weight helps to avoid chattering.

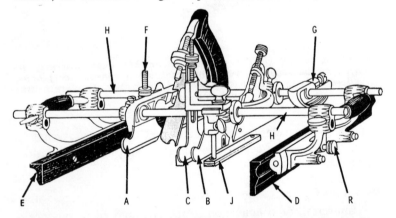

Fig. 11. A universal plane. This plane may be used for several different operations, such as plowing, dadoing, rabbeting, beading, reeding, fluting, rounding, hollowing, slitting, chamfering, and general moulding. The plane consists of the main stock (A), which carries the cutter, cutter adjustment, and cutter bolt, slitting tool, depth gauge, and handle; the sliding section (B); the auxiliary center bottom (C); the fences (D and E); the gauges (F and J); the arms for carrying the fences (H); and the cam rest (G).

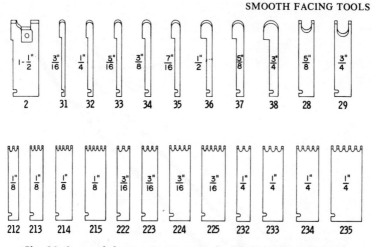

Fig. 12. Some of the many cutters used with the universal plane.

The thin plane iron is normally satisfactory when the plane is properly constructed so that firm support is given the cutter over a considerable portion of its length. It also has the slight advantage of requiring less grinding.

For the first, or roughing, cut with the jack plane, the cutting edge is ground slightly curved (convex), as shown in Fig. 15, because since it is used for heavy work, it removes thick shavings; if the cutter were ground straight, the plane would cut a rectangular channel from which the wood must be torn as well as cut, as shown in Fig. 16A. Moreover, such a shaving would probably stick fast in the throat of the plane or require undue force to push the plane. Compare this with the shaving taken from the fully curved cutting edge of the jack plane, as shown in Fig. 16B.

When a full set of planes is available, the fore plane should have some curvature to the cutting edge. In this case, the process of transforming the grooved surface produced by the jack plane to a flat surface is accomplished in three operations, using the jack, fore, and jointer plane, as shown in Fig. 17. The cutting edge of the jointer and smoothing plane irons are made straight with rounded corners, as shown in Fig. 15. Because this type of plane iron makes an extremely fine cut, the groove caused by the removal of so delicate a shaving is sufficiently blended with the general work by the rounded corners of the iron.

183

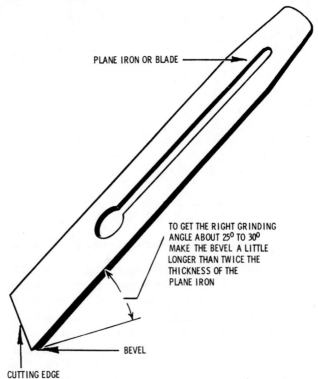

PLANE IRON OR BLADE

TO GET THE RIGHT GRINDING
ANGLE ABOUT 25° TO 30°
MAKE THE BEVEL A LITTLE
LONGER THAN TWICE THE
THICKNESS OF THE
PLANE IRON

BEVEL

CUTTING EDGE

Fig. 13. A single plane iron.

Bevel of the Cutting Edge

Many of the complaints concerning poorly cutting plane irons are due to improper plane-iron grinding. The bevel should always be ground at an angle of 25°, which means that it must be twice as long as the cutter is thick. If the bevel is too long, the plane will jump and chatter; if it is too short, it will not cut. It is a good rule, perhaps, to have a long thin bevel for softwood and a 25° bevel for the hardwoods, although cross-grained timber requires a short bevel.

Double Irons

The term "double iron" means a plane iron equipped with a supplementary iron called a "cap." The object of the cap is to

184

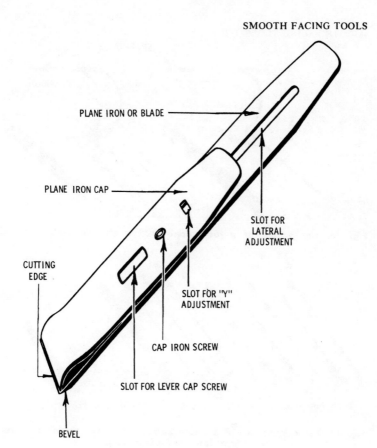

PLANE IRON OR BLADE

PLANE IRON CAP

SLOT FOR LATERAL ADJUSTMENT

CUTTING EDGE

SLOT FOR "Y" ADJUSTMENT

CAP IRON SCREW

SLOT FOR LEVER CAP SCREW

BEVEL

Fig. 14. A double plane iron.

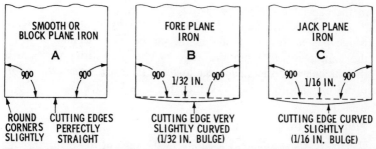

SMOOTH OR BLOCK PLANE IRON

A

90° 90°

ROUND CORNERS SLIGHTLY

CUTTING EDGES PERFECTLY STRAIGHT

FORE PLANE IRON

B

90° 1/32 IN. 90°

CUTTING EDGE VERY SLIGHTLY CURVED (1/32 IN. BULGE)

JACK PLANE IRON

C

90° 1/16 IN. 90°

CUTTING EDGE CURVED SLIGHTLY (1/16 IN. BULGE)

Fig. 15. Cutting edges for common plane irons. These edges should be straight on smooth and block plane irons and just slightly curved on jack and fore plane irons.

185

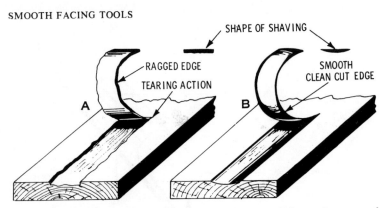

Fig. 16. *The actions of a jack plane with a straight and a curved cutter.*

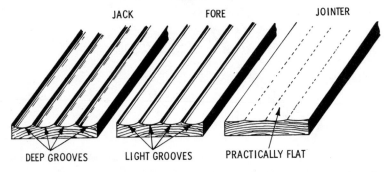

Fig. 17. *The appearance of a board surface after having been planed with jack, fore, and jointer planes.*

break the shaving as soon as possible after it is cut. The action of the cap is shown in Fig. 18B.

The cap is attached to the cutting iron by tightening a screw which passes through a slot in the cutter. The distance at which the cap is placed from the edge of the plane iron varies with the thickness of the shaving; allow $\frac{1}{32}$ inch for a smooth or fore plane and approximately $\frac{1}{8}$ inch for a jack plane.

Plane Mouth

This is the rectangular opening in the face of the plane through which the cutter projects and, in operation, through which the shavings pass. The width of the mouth has an important bearing

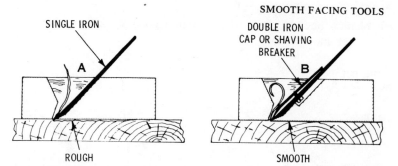

SINGLE IRON

A

ROUGH

DOUBLE IRON
CAP OR SHAVING
BREAKER

B

SMOOTH

Fig. 18. Action of single and double plane irons. The single iron cuts satisfactorily only when the grain is favorable, but when the grain varies from the line of cut, the shavings run up the iron, thereby leaving a rather rough surface.

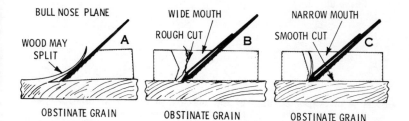

BULL NOSE PLANE

WOOD MAY
SPLIT

A

OBSTINATE GRAIN

WIDE MOUTH

ROUGH CUT

B

OBSTINATE GRAIN

NARROW MOUTH

SMOOTH CUT

C

OBSTINATE GRAIN

Fig. 19. Influence of mouth width. The bull-nose plane, may be regarded as an ordinary plane with a mouth of infinite width; since there is nothing in front of the cutter to hold down the wood, a splitting action is possible with obstinate grain. B and C show the results obtained with wide- and narrow-mouth planes.

on the proper working of the plane. That portion of the plane face in front of the mouth prevents the wood from rising in the form of a shaving before it reaches the mouth. If there was no face in front of the cutter, as in the case of a bull-nose plane, there would be nothing to hold down the wood in advance of the cutter, and the shaving would not be broken. In obstinate grain, the work will be rougher, and a splitting instead of a cutting action may result, as illustrated in Fig. 19. Accordingly, the wider the mouth, the less frequently the shaving will be broken and, in obstinate grain, the rougher the work.

How to Use a Plane

In order to obtain satisfactory results with planes, it is necessary to know not only the proper method of handling the tool

187

in planing but also how to put it into good working condition. The user must know:

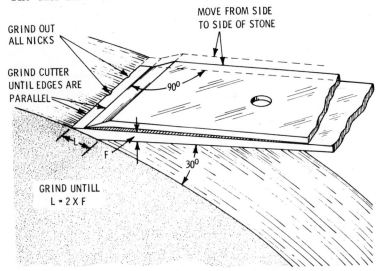

Fig. 20. The proper position of the plane iron on the grinding wheel. Note the conditions which must be fulfilled to grind properly.

1. How to sharpen the cutter.
2. How to adjust the cutter.
3. How to plane.

Sharpening Planes—This involves two operations, grinding and whetting. When grinding, the cutter must be ground perfectly square; that is, the cutting edge must be at right angles to the side. Enough metal must be removed to remove all nicks in the cutting edge. Before grinding, loosen the cap and set it back approximately ⅛ inch from the edge; it will serve as a guide by which to square the edge.

The cutter should be held firmly on the grinding wheel at the proper angle and should be moved continually from side to side to prevent wearing the stone out of true. Grind on the bevel side only. The bevel angle should be approximately 30°. This angle is attained when the length of the bevel is twice the thickness of the cutter. Fig. 20 shows the proper position of the cutter

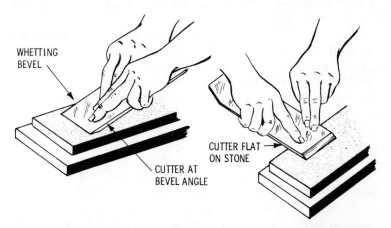

Fig. 21. *The method of whetting the plane iron on an oilstone after grinding. Grasp the plane iron firmly in the right hand with the palm down, pressing down with the left hand near the cutting end to provide rubbing pressure. Rub back and forth along the length of the stone. After whetting the bevel side, turn the plane iron over and hold it perfectly flat on the stone; give it two or three strokes to remove any wire edge.*

on the grinding wheel. The edge should be ground to one of the forms shown in Fig. 15, depending on the type of plane and the requirements of the work to be done.

After grinding, the cutter will have a wire edge; that is, the coarse grit of the grinder will always leave the edge comparatively coarse or rough, and the edge will not be as keen as it should be to cut smoothly. This wire edge is removed by the aid of an oilstone, as shown in Fig. 21.

In the case of a double iron, the cap should be kept with a fine but not a cutting edge. The cap must be made to fit the face of the cutter accurately; if it does not fit precisely, the plane will quickly "choke" with shavings because of the shavings driven between the two irons. This is an extremely important point, and it should be noted that even a minute opening between the plane irons will allow the shavings to drive in and choke the plane.

Adjusting the Cutter—After sharpening the cap of a double iron, position the screw in the slot; tighten the screw lightly on the cap to within ¼ inch of the cutting edge, then tighten the screw. Finish the setting by driving the cap up to its final position, tapping lightly on the setscrew.

The "set" of the iron is the amount of cutter face exposed below the edge of the cap. The plane iron is said to be set coarse or fine according to the amount of cutter face exposed. The set, therefore, regulates the thickness of the shavings and is varied according to the nature and kind of wood to be planed. For soft woods, the set should be: $\frac{1}{2}$ inch for the jack plane, $\frac{1}{16}$ inch for the jointer plane, and $\frac{1}{32}$ inch for the smoothing plane. If the wood is hard or cross-grained, allow for approximately one-half of these settings.

How to Plane—Satisfactory results in the use of a plane depend largely on the plane being in perfect condition and properly adjusted with respect to set and depth of cut to suit the kind of wood to be planed. The first point to learn is the correct way of holding the plane. Do not allow the plane to drop over the end of the board at either end of the stroke. Before planing, examine the board with respect to the grain, and turn the board so as to take advantage of the grain. On the return stroke, lift the back

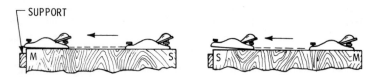

Fig. 22. The method of using the block plane across the grain. The cut must not be taken across the entire length of the board to prevent the board from splitting. Lift the plane before the cutter runs off the edge. Take several strokes with the board in position MS; then reverse the board to position SM and continue planing.

of the plane slightly so that the cutter does not rub against the wood, thus preventing the cutter from being quickly dulled.

When planing a narrow surface, let the fingers project below the plane, and use the side of the board as a guide to keep the plane on the work. If the plane chokes with shavings, look for and repair the cause instead of just removing them. Remove the iron and carefully examine the edge of the cap. This must be a perfect fit or there will be continual trouble.

To plane a long surface, such as a long board, begin at the right-hand end. Take a few strokes, then step forward and take the same number of strokes, progressing this way until the entire surface is passed over. To preserve the face of the plane, apply a few drops of oil occasionally to the cutter.

When cutting across the grain with a block plane, the cut should not be taken entirely across, but the plane should be lifted before the cutter reaches the edge of the board; if this precaution is not taken, the wood will split at the edge. After taking a few strokes, reverse the board and continue as directed, as shown in Fig. 22.

SCRAPERS

The term "scraper" usually signifies a piece of steel plate of approximately the thickness and hardness of a saw. There are several types of scrapers:

1. Unmounted.
2. Handle scraper.
3. Scraper plane.

The unmounted scraper is simply a rectangular steel blade, whose cutting edges are formed by a surface which is at right angles to the sides. Quicker cutting is secured with the cutting

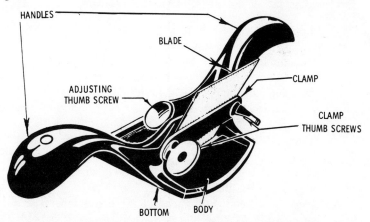

Fig. 23. A typical cabinet scraper. In operation, the blade springs backward, thereby opening the mouth and allowing the shavings to pass through. As soon as the working pressure is released, the blade springs back to its normal position.

191

edge at a more acute angle, but more labor is required to keep it sharp. The cutting edge is sharpened by filing or grinding. For smooth work, the roughness of the edge may be removed by an oilstone, but the rougher edge will cut faster.

When cutting, the scraper is inclined slightly forward and is more conveniently held when provided with a handle or mounted like a plane. A typical scraper is shown in Fig. 23. The method of using a hand scraper is illustrated in Fig. 24.

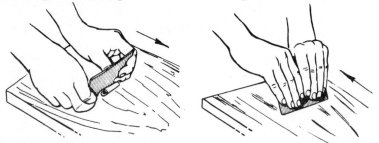

Fig. 24. A hand scraper and its method of use. In operation, the hand scraper is pushed or pulled, thus removing surface irregularities in the work; when properly used, it provides the work with a smooth and glossy finish.

To sharpen the scraper blade for heavy work so as to avoid too frequent use of the oilstone, use the following procedure:

1. File the cutter to a keen edge; remove the wire edge with a coarse, medium oilstone.
2. Hold the burnisher in both hands, and turn the edge.
3. Begin with light pressure, and hold the steel at nearly the same angle as the file was held in filing.
4. Bear harder for each successive stroke, and allow the tool to come a little nearer level each time; finish with the tool at an angle of approximately 60° from the face of the blade.
5. Be sure that the steel never comes down squarely on the fine edge, since that will ruin it.
6. Keep the edge slightly ahead of the cutter face. The object is to get a hook edge that is sharp.

By filing and honing the scraper blade square, two usable cutting edges are produced. Remember, the edges must be *smooth and sharp* before burnishing. Some workmen prefer to sharpen the cutter with a bevel having only one cutting edge. With this type of blade, a heavy cutting burr can be turned, and, for heavy use in handled scrapers or planes, such as on floors, it works better; sometimes the edge can be reburnished and sharpened without refiling and honing. All good burnishers have a blunt point; two typical burnishers are shown in Fig. 25. Run the point of the burnisher along under the edge, partially turning it back, and

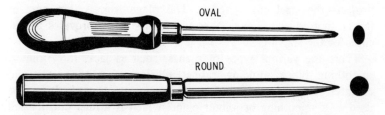

Fig. 25. *Scraper steels or burnishers. This tool is used to turn the cutting edge of a scraper after filing and honing. "Turning" means pushing the particles of steel which form the corner over so that they will form a wire edge that will stand at an angle with the sides of the scraper.*

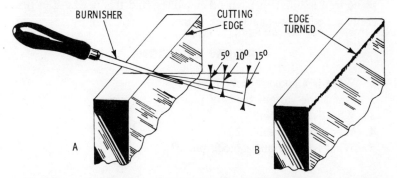

Fig. 26. *The application of a burnisher to turn the edge of a scraper after filing and honing. The edge is usually turned in two or three strokes with the burnisher at angles of 5°, 10°, and finally 15°, as shown in A. The appearance of the finished edge, completely turned, is shown in B.*

alternately burnish it down again, as shown in Fig. 26. The highly skilled workman, with care, can do this two or more times.

It is impossible to give unfailing, inflexible instructions. Consistently good sharpening of scrapers comes only with practice and experience. There is no surer indication of an experienced carpenter than his skill in sharpening scrapers.

SUMMARY

The woodworking plane consists essentially of a smooth-soled stock of wood or iron, from the underside (or face) of which projects the steel cutting edge. The plane iron is that part of the cutting edge or knife. A section in the front provides an escape for the shavings.

There are various types of planes, such as jack, fore, jointer, smoothing, block, molding, rabbet, fillister, grooving, router, and chamfer. The jack plane is for very heavy rough work. It can be used as a smoothing or jointer plane, which is a finer cut, if used properly.

Block planes are used primarily as a one-hand plane and almost exclusively for planing across the grain. The angle of the plane iron is different from other planes, and this is why it is generally called a low-angle block plane.

REVIEW QUESTIONS

1. What is a spokeshave?
2. Name the various types of planes.
3. What is a scraper?
4. What is a plane iron?
5. What is a low-angle block plane?

CHAPTER 16

Boring Tools

There are several kinds of boring tools; each class is adapted to meet special working conditions, such as:

1. Punching.
2. Boring.
3. Drilling.
4. Countersinking.
5. Enlarging.

The various kinds of tools used for these operations are brad awls, gimlets and augers, drills, hollow augers and spur pointers, countersinks, and reamers. These tools are called bits when provided with a shank instead of a handle for use with a brace or for use in an electric drill or drill press.

SCRATCH AWL

An awl is a pointed tool that is used for small starting holes for screws and nails. It can also be used to accurately scribe a line. Fig. 1 shows the scratch awl.

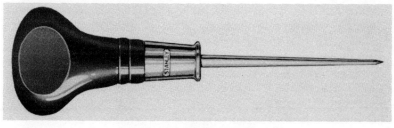

Fig. 1. A typical scratch awl.

AUGERS

Augers are used for boring holes from ¼ inch to 2 inches. The sizes of auger and Forstner bits are listed in 16ths; thus, a 2-inch auger is listed as size 32. Fig. 2 illustrates a comparison between auger, Forstner, and twist bits. When made with a shank for use in a brace, this style of auger is commonly called a bit, as shown in Fig. 3.

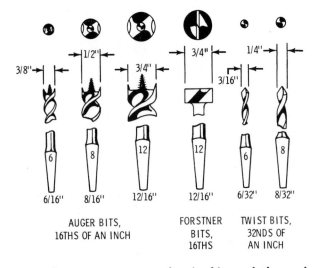

AUGER BITS, 16THS OF AN INCH

FORSTNER BITS, 16THS

TWIST BITS, 32NDS OF AN INCH

Fig. 2. Typical auger, Forstner, and twist bits and the methods of marking their size.

Because of the enormous variety of bits on the market, it is difficult to select the one best adapted for a given purpose. For accurate boring, for rapid boring, for rough boring, the bit adapted for the purpose must be used to get the proper result. A small example of this fact is illustrated in Fig. 4, which shows an enlarged view of just two common styles of wood bits. One has a screw point, and the other has a brad, or diamond, point. Note also that one style has a solid center with a single spiral running around it, while the other is a double-spiral twist bit.

It is not generally understood how important a part the screw thread plays in boring. The terms "coarse" and "fine," as applied to a screw thread, are relative and may be applied to either single-

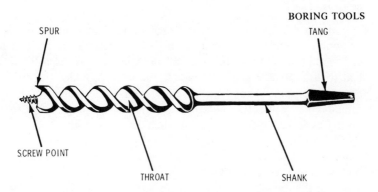

Fig. 3. A typical auger bit illustrating its component parts.

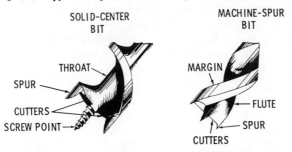

Fig. 4. Two common types of wood bits. The cutting edges of these bits are quite similar. The opening between the spiral is called the throat, but in some styles of double-twist bits, it is called the "flute." Both terms are used alternately and mean the same thing.

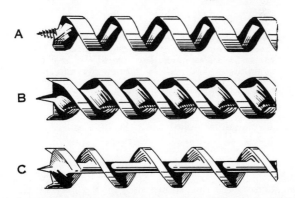

Fig. 5. Typical auger-bit styles. The auger bits shown represent the hollow-spiral bit (single twist) (A), the double-spur bit (fluted) (B), and the double-spur bit (solid center) (C), respectively.

thread or double-thread screws. The bit having a given number of double threads to the inch, provided the cutters are pitched to correspond with those threads, will bore just as fast as a bit with half that number of single threads to the inch, provided the cutters are of the same pitch. If the cutters have less pitch than the threads, they will act as a stop gauge, thereby keeping the bit from boring as fast as it would without such an obstruction.

Fig. 5 shows a hollow-spiral and two double-spur bits. The hollow-spiral bit has a screw point and only one cutting edge; its hollow center permits the easy passage of wood chips, thereby allowing this type of bit to cut faster, especially when boring deep holes. The expansive bit, shown in Fig. 6, is so called because it may be set for various diameter holes within its capacity, thereby taking the place of many large bits.

Another excellent bit to use when boring large holes is the multispur type, shown in Fig. 7A; although this bit is relatively low in price, it does not readily bore holes of various diameters. A double-spur, twist-drill bit is shown in Fig. 7 B; this is one of the cleanest and fastest cutting types of wood bits on the market. The flat-center bit, Fig. 7 C, has only one cutting edge and is used

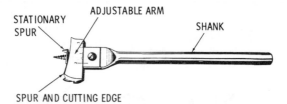

STATIONARY SPUR ADJUSTABLE ARM SHANK

SPUR AND CUTTING EDGE

Fig. 6. A typical expansive auger bit with an adjusting screw. The expansive bit obsoletes the necessity for many large bits. The cutter may be adjusted for various size holes. The size of the hole to be cut may be reduced or enlarged 1/8 inch by turning the adjusting screw one complete revolution in the direction desired. Test for the correct size setting on a piece of waste wood before boring the hole in the wood being worked.

for boring large shallow holes. Both the hole and the countersink can be cut in one operation with the countersink bit illustrated in Fig. 7D.

If it is necessary to drill holes to an exact depth, an adjustable bit gauge, such as the one shown in Fig. 8 , may be used. This tool is simply a clamp that can be securely attached to any standard-size wood bit by means of two wing nuts, as illustrated.

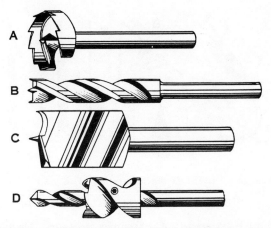

Fig. 7. Special wood-bit styles. The bits represented are a multispur bit (A), a double-spur bit (B), a center bit (C), and a countersink bit (D). Bits not equipped with feed-screw points are usually meant to be used in the drill press, whereas bits equipped with screw points are for use with the hand brace.

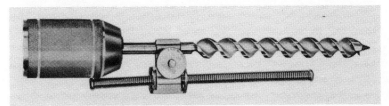

Fig. 8. A typical depth gauge and its use. An adjustable bit gauge of this type may be used to regulate the depth of the holes to be bored.

Among other types of bits frequently used in woodworking shops are router bits, end cutters (for cutting rosettes, rounding, and shaping), and a variety of other designs. Several common types of bit styles are shown in Fig. 9 .

It should be clearly understood that the double-thread bit is intended for softwood, and the single-thread bit is intended for hardwood. The single-thread bit will not clog up as readily as the double-thread type; if the double-thread bit were left coarse enough so as not to clog, it would make the bit turn too hard.

To sharpen the spur of an auger, hold the bit in the left hand with the twist resting on the edge of the bench. Turn the bit

199

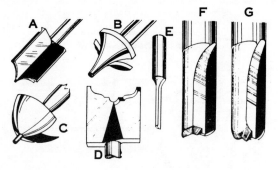

Fig. 9. Special bit styles. In the illustration, A, B, and C represent end cutters; D shows a rosette cutter available in many different patterns; E, F, and G are different types of router bits.

around until the spur to be sharpened faces up. File the side of the spur next to the screw, carefully keeping the original bevel. File lightly until a fine burr shows on the outside, which is carefully removed by a slight stroke with a file; the result is a fine cutting edge.

To sharpen the cutter, hold the bit firmly in the left hand with the worn point down on the edge of the bench, slanted away from the hand with which you file. File from the inside back, and be careful to preserve the original bevel; take off the burr or rough edge. Never sharpen the outside of the spur.

It is rarely necessary or advisable to sharpen the worm; however, it may often be improved if it is battered by using a three-cornered file that is carefully manipulated; use a size that fits the thread. A half-round file is best for the lip and, with careful handling, may be used for the spur. Special auger-bit files are available for this purpose.

TWIST DRILLS

In addition to augers and gimlets, a carpenter should possess a set of twist drills. These tools are used for drilling small holes where the ordinary auger or gimlet would probably split the wood. They come either with square shanks for use with bit braces or with straight shanks for use with breast or electric drills, as shown in Fig. 10. These drills are available in standard sizes from $\frac{1}{16}$ to $\frac{5}{8}$ inch or more, varying in size by 32nds of an inch.

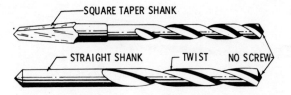

Fig. 10. A bit stock twist drill for use with a brace and a straight-shank twist drill for use in the drill press.

A twist drill differs from an auger or gimlet in that it has no screw and has a less acute cutting angle of the lip; therefore, there is no tendency to split the wood, since the tool does not pull itself in by a taper screw but enters by external pressure.

For many operations, especially where the smaller drills are used, as in drilling nail holes through boat ribs and planking, a geared breast drill is preferable to a brace.

COUNTERSINKS

Sometimes it is necessary to make a conical enlargement of a hole at the surface of the wood. This operation is performed by a

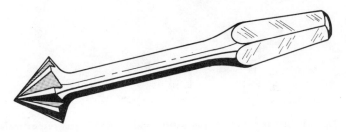

Fig. 11. A typical rose countersink.

bit tool called a countersink, which may be used in a hand brace, an electric drill, or a drill press. A typical countersink is shown in Fig. 11.

REAMERS

A reamer is a long tapered cutting tool that is used for enlarging holes. Although this type of tool is used chiefly by ma-

Fig. 12. An octagonal-type reamer.

chinists, there are frequent occasions in carpentry when a reamer may be employed advantageously, such as for enlarging holes in hinges when they are too small for the screws on hand, etc. Fig. 12 shows a desirable type of octagonal reamer with a square shank for use with a brace.

Fig. 13. A hand drill, a breast drill, and a hand brace, respectively.

HAND DRILLS, BREAST DRILLS, AND BRACES

The hand drill, breast drill, and brace, Fig. 13, are the conventional hand tools used by the carpenter for holding and turning bits. The hand drill is used for rapid drilling of small holes. The breast drill is similar in design to the hand drill but is equipped with a breastplate to facilitate the application of additional pressure on the drill when required. The brace differs from the hand and

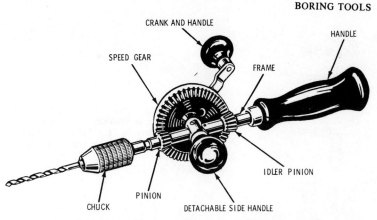

Fig. 14. The component parts of a typical hand drill.

breast drills mainly in that with the brace, the turning movement is applied directly to the bit by means of the handle swing, where-as the hand and breast drills are equipped with a gear-pinion ar-rangement for turning the drill. The component parts of the hand drill and the brace are identified in Figs. 14 and 15 , respectively,

Satisfactory results in the use of boring tools are only obtained with practice and the use of good tools, each suitable for the particular job assigned to it. The work should be properly laid out, and the hole should be clearly marked. To bore a vertical hole, hold the brace and bit perpendicular to the surface of the work, as shown in Fig. 16 . Compare the direction of the bit to the closest straightedge or to the sides of the vise. A try square may also be held near the bit to be certain of the true vertical position.

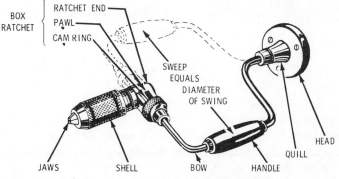

Fig. 15. The component parts of a typical hand brace.

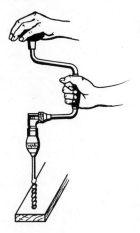

Fig. 16. **The proper method of boring a vertical hole with the hand brace. The bit must be perpendicular to the work surface.**

To bore a horizontal hole, hold the head of the brace cupped in the left hand against the stomach, with the thumb and forefinger around the quill, as shown in Fig. 17. To bore through the wood without splintering the second face, stop when the screw point reaches the other side, and finish the hole from this side. When boring with an expansive bit, it is best to clamp a piece of scrap wood to the second face and bore straight through.

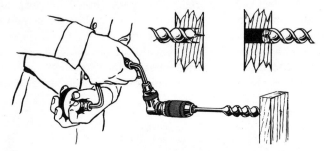

Fig. 17. The correct method of boring holes horizontally to prevent splitting and splintering.

Frequently, restricted working quarters make it necessary to use the ratchet device of the hand brace. The ratchet brace is indispensable when boring a hole in a corner or when a projecting object prevents the user from making a full turn with the handle. To actuate the ratchet, turn the cam ring. Turning the cam ring to the right will allow the bit to turn clockwise and give a ratchet action when the handle is turned left; turning the cam ring to the left will reverse this action.

SUMMARY

Various kinds of boring tools are used in woodworking shops. Tools like punches, drills, countersinks, enlargers, and boring implements are used everyday in various operations. These tools are generally called bits when provided with a shank instead of a handle for use with a brace or for use in an electric drill.

There are various types and sizes of augers. Some augers have a screw point, and others have a brad, or diamond point. In Chapter 16, Fig. 5 illustrates three different auger styles which are the most popular.

Twist drills are used for drilling small holes. They are designed with square shanks for brace or straight shank for electric drills. The twist drill differs from an auger in that it has no screw and has a less acute cutting angle of the lip; therefore there is less danger of splitting the wood.

REVIEW QUESTIONS

1. What is a gimlet?
2. Explain the auger and its uses.
3. What is the difference between the flute and throat on a wood bit or auger?
4. What is the spur of a wood bit?
5. What is a reamer and how is it used?

CHAPTER 17

Fastening Tools

The term "fastening tools" means those hand tools that are used in the operation of securing or joining various parts of the work that must be fastened together with nails, tacks, screws, bolts, etc. The tools used for these operations comprise the various hammers, screwdrivers, and wrenches.

HAMMERS

The hammer is the all-important tool in carpentry, and there are numerous types to meet the varied conditions of use. All hammers worthy of the name are made of the best steel available and are carefully forged, hardened, and tempered.

The shapes of the claws of hammers vary slightly in the products of different manufacturers, though they may all be called curved-claw hammers. Carpenters often develop a preference. The straight-claw, or ripping, hammer is not as popular as the curved-claw hammer, shown in Fig. 1, with good workmen, because it does not grip nails for withdrawal quite so readily, and they cannot be withdrawn easily. The shape of the poll is immaterial, but the octagon-shaped poll seems to be most popular. All good hammers have slightly rounded faces, thereby making it possible to drive a nail head down flush with the wood without unduly marking the wood.

Several types of hammers are presently available, as shown in Figs. 2 to 5. One popular type is forged (head and handle) of a single piece of drop-forged steel with the grip built-up of leather or neoprene. Fiberglass-handle hammers are becoming very popular because of their greater strength, finer balance, and light weight.

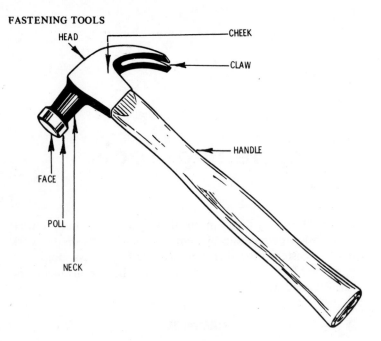

Fig. 1. A typical bell-faced nail hammer.

Courtesy Vaughan and Bushnell Mfg. Co.

Fig. 2. Full polished octagon neck, round face, air-cushioned neoprene grip hammer. This hammer is forged steel from head to toe with a hickory plug in the head to absorb shock.

The handle is made of polyester resin reinforced by continuous fiberglass in parallel form for greatest possible strength. They are available in 13-, 16-, and 20-oz. nail types, as well as 16- and 20-oz. ripping types, with nonslip neoprene grips.

Despite the obvious advantages in these types of hammers, some of the best workmen still prefer the old standard adze-eye hammer with a handle of springy hickory. True, such handles are rather easily sprung, and are often broken, but the "feel" of such ham-

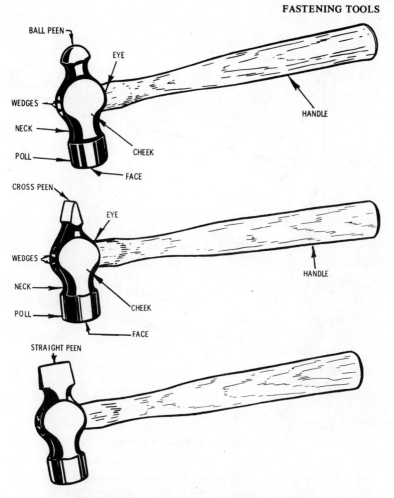

Fig. 3. Ball-, cross-, and straight-peen hammers.

mers is certainly different from those with metal handles. Unquestionably though, anyone can become accustomed to the metalhandled tools in time.

It is not a good policy for the beginner to change to a lighter hammer when doing trim work. He will usually be slow enough to acquire good control of one hammer. Builders and other workmen who do an excessive amount of heavy spiking often use a 2-pound hammer, sometimes with a cut face. Of course, such

209

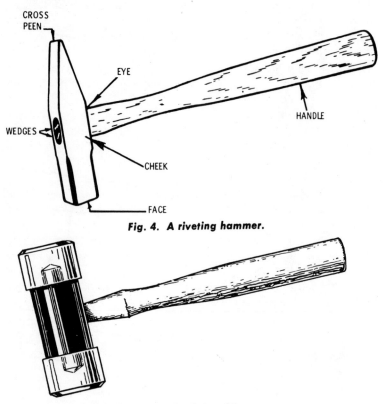

Fig. 4. A riveting hammer.

Fig. 5. A typical soft-faced hammer.

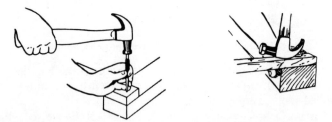

Fig. 6. The method of setting the nail below the surface of the wood and the use of the claws when pulling nails are also shown.

hammers are of no use for anything else. The beginner should select a good hammer from an established and recognized manufacturer. The advice of an experienced foreman or an old carpenter is also extremely desirable. The novice will bend a sufficient number of nails anyway with the best of hammers until he acquires skill in using this, his most indispensable tool.

When using a hammer, the handle should be grasped at a short distance from the end, and a few sharp blows rather than many light ones should be given, as shown in Fig. 6. Keep the hand and wrist level with the nail head so that the hammer will hit the nail squarely on the head instead of at an angle. Failure to do this is the reason for the difficulty so often experienced in driving nails straight.

The face must be free from grease or dirt to drive a nail straight; therefore, rub the face of the hammer frequently on wood. Hammers are designed to drive nails and not to hit wood (or fingers); thus, when starting tap gently while the nail is guided with the fingers and finish with a nail set. Hammers vary in size from 5 to 20 ounces. The bench worker usually employs a hammer weighing 14 to 16 ounces.

It is always a good idea to protect eyes and face from dust and flying particles when using a hammer. Many safety goggles are designed to be worn over personal glasses, with large clear-vision lenses and air holes for maximum ventilation, as shown in Fig. 7.

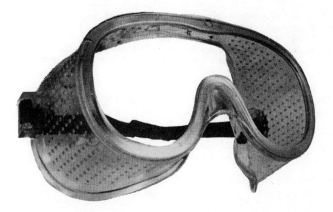

Courtesy Vaughan and Bushnell Mfg. Co.

Fig. 7. Safety goggles for use with striking tools.

SCREWDRIVERS

A screwdriver is quite similar to a chisel and usually differs only in the working end, which is blunt. There are few screwdrivers that have a correctly shaped end. Usually the sides which enter the slot in the screw are tapered. This is done so that the end of the screwdriver will fit into screw slots of widely varying sizes, as shown in Fig. 8.

When using a screwdriver with a tapered blade tip, a force is set up due to the taper which tends to push the end of the tool out of the slot. Therefore, it is better to have several sizes with properly shaped parallel sides than to depend on one size with tapered sides for all sizes of screws.

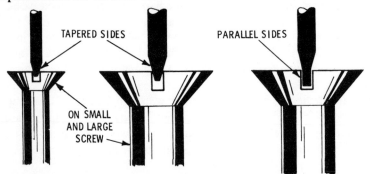

Fig. 8. The end of a screwdriver should be shaped so that its sides are parallel. A screwdriver whose end is tapered can be used, but considerable downward pressure must be exerted to prevent the screwdriver from rising out of the screw slot. With parallel sides, there is no tendency for the screwdriver to rise, no matter how much turning force is exerted.

There are two general classes of screwdrivers—the plain and the so-called automatic. Fig. 9 shows a typical plain screwdriver. The operation of driving a screw with a plain screwdriver consists of giving it a series of half turns.

Where a number of screws are to be tightened, time can be saved by using a screwdriver bit which is used with a brace in the same manner as an auger bit. The quickest method of driving a screw is by means of the so-called automatic ratchet-type screwdriver, shown in Fig. 10. The advantage of this type over the plain screwdriver is that instead of grasping and releasing the handle

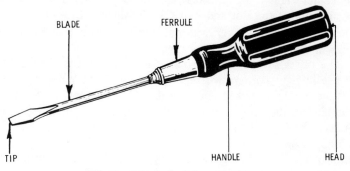

BLADE　　　　　FERRULE

TIP　　　　　　　　　　　　HANDLE　　　　HEAD

Fig. 9. A typical plain screwdriver.

from 25 to 30 times in turning a screw home, it is grasped once and with two or three back-and-forth strokes, the screw is driven home, thus saving labor and time. The screwdriver drives or withdraws screws according to the position of the ratchet shifter, because pressure on the handle causes the spindle and tip to rotate. The ratchet shifter can also lock the screwdriver into a rigid unit like a conventional plain screwdriver.

Special ratchet-type screwdrivers may be obtained with spirals of different angles to suit working conditions, such as: a 40° spiral for rapidly driving small screws, a 30° spiral for general work, and a 20° spiral for driving large screws in hardwood.

The Phillips screwdriver, Fig. 11, is constructed with a specially shaped blade tip to fit Phillips cross-slot screws; the heads of these screws have two perpendicular slots that intersect each other in

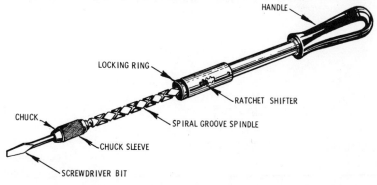

HANDLE

LOCKING RING

RATCHET SHIFTER

CHUCK

SPIRAL GROOVE SPINDLE

CHUCK SLEEVE

SCREWDRIVER BIT

Fig. 10. | A spiral-ratchet screwdriver.

213

the center. This design checks the tendency of some screwdrivers to slide out of the slot and on to the finished surface of the work. The Phillips screwdriver will not slip and burr the end of the screw if the proper size tool is used.

The offset screwdriver, also shown in Fig. 11, is a handy tool to use in tight corners where working room is limited. It usually has one blade forged in line with the shank, or handle, but this is not always the case; the other blade is at a right angle to the shank. The ends of the screwdriver can be changed with each turn, thereby working the screw into or out of its hole. This type of screwdriver is normally used only when the screw location is such that it prohibits the use of a plain or spiral-ratchet screwdriver. The offset screwdriver is also available in the ratchet form.

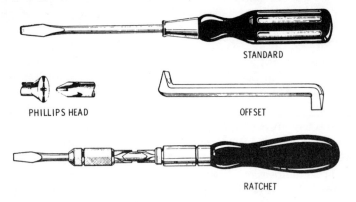

STANDARD

PHILLIPS HEAD   OFFSET

RATCHET

Fig. 11. A typical assortment of screwdrivers.

WRENCHES

There are presently available on the market an undue multiplicity of wrenches of many kinds and patterns for every conceivable use. The wrench, though it may not be so considered, can be a somewhat dangerous tool, especially when great force is applied to start an obstinate nut. Often under such conditions, the jaws slip off the nut, thereby resulting in injury to the workman by violent contact with some metal part.

There are three general classes of wrenches:

214

1. Plain.
2. Adjustable.
3. Socket.

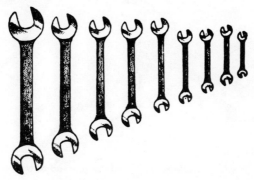

Fig. 12. Open-end wrenches.

Plain, or open-end, wrenches are of the solid nonadjustable type with fixed openings at each end, as shown in Fig. 12. A conventional set of wrenches for the carpenter's use contains between eight and twelve wrenches ranging in size from $\frac{5}{16}$ to 1 inch in jaw width.

Another form of plain wrench is the box-end wrench; these are so called because they "box," or completely surround, the nut or bolt head. Their usefulness is based to some extent on their ability to operate in close quarters. Because of their unique design, there is little chance of the wrench slipping off the nut. As shown in Fig. 13, twelve notches are arranged in a circle around the inside of the "box." This twelve-point wrench can be used to continuously

Fig. 13. A typical box-end wrench.

loosen or tighten a nut with a handle rotation of only 15°, as compared to the 60° swing required by the open-end wrench if it is reversed after each swing. A combination wrench employs a box-end wrench at one end and an open-end wrench at the other;

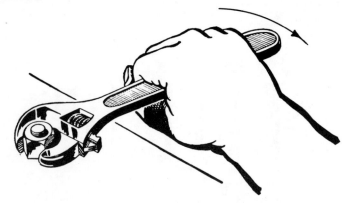

Fig. 14. The adjustable wrench and the method of tightening a nut.

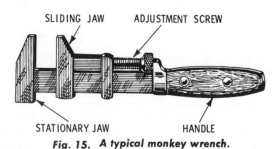

Fig. 15. A typical monkey wrench.

both ends are usually the same size, although this is not always the case.

Adjustable wrenches are somewhat similar to open-end wrenches; the primary difference is that one jaw is adjustable, as shown in Fig. 14. The angle of the opening with respect to the handle on an adjustable wrench is 22.5°. A spiral worm adjustment in the handle permits the width of the jaws to be varied from zero to ½ inch or more, depending on the size of the wrench. Always place the wrench on the nut so that the pulling force is applied against the *stationary* jaw. Tighten the adjusting screw so that the jaws fit the nut snugly.

The wrench most commonly found in a carpenter's toolbox is probably the old-fashioned monkey wrench, shown in Fig. 15. For most of his uses, this wrench fills its purpose well. Fit the jaws of the wrench snugly on the nut to be turned; it is usually advisable to turn *toward* the screw side of the handle, since the wrench is

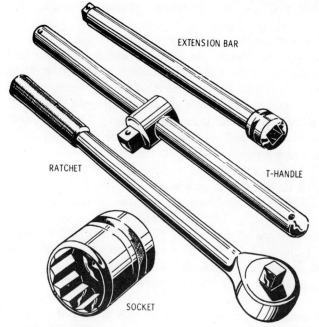

Fig. 16. *The socket wrench and the various handles used with it.*

then not so likely to slip off the nut. Do not try to turn the nut or bolt with the tips of the jaws, or slip a piece of pipe over the handle to increase leverage; a stillson pipe wrench may be able to take it, but an ordinary monkey wrench won't. Keep the adjustment screw well oiled.

Socket wrenches, as shown in Fig. 16 , are so called because the wrench is made in two or more parts—a socket that fits the bolt or nut and a detachable handle or lever to fit the socket. The modern socket wrench is usually equipped with a ratchet mechanism similar to that used in certain hand braces for boring holes in tight corners. This unit permits the wrench to be used in locations where the handle-travel space is limited; its use also makes it unnecessary to shift the socket on the nut or bolt for each pull on the handle.

SUMMARY

Fastening tools include various hammers, screwdrivers, and wrenches. These tools are used for securing or joining various parts of wood or other materials together with nails, screws, or bolts.

The hammer is a very simple striking tool and is made in numerous sizes and shapes to meet various conditions. All hammers worthy of the name are made of the best steel, carefully forged, hardened, and tempered. There are various groups or classifications, such as nail or claw, ball peen, soft face, cross and straight peen, and tinners and riveting. Hammer handles are usually made of hickory and carefully balanced to deliver maximum striking power. Hammers, regardless of their classification, are sized according to the weight of the head without the handle, such as 10, 13, 16, 20 oz., etc.

Screwdrivers are designed to insert or remove screws from various materials. There are several classes of screwdrivers used in the average shop, such as plain, Phillips, offset, and ratchet. The plain screwdriver is very similar to a chisel except it has a blunt end. The Phillips screwdriver is made with a specially shaped blade tip that fits Phillips cross-slot screws.

The wrench is a tool for tightening or loosening bolts and nuts used in the assembly of numerous articles of wood or other material. The majority of nuts and bolt heads are hexagonal (six-sided), although other shapes are sometimes encountered. The wrench is designed to grip these nuts and bolt heads and turn them by means of lever action exerted at the handle. Various types manufactured are open end, monkey, adjustable, box end, socket, and combination.

REVIEW QUESTIONS

1. What is a Phillips screwdriver?
2. How are hammers sized for a particular job?
3. Name the various type wrenches manufactured.
4. Name the various types of hammers manufactured.
5. What are the requirements for a good screwdriver?

CHAPTER 18

Sharpening Tools

Too much cannot be written for the amateur on the subject of sharpening tools and the methods of sharpening. The tools used for sharpening by the carpenter, in addition to files which have already been described, are:

1. Grinding wheels.
2. Oilstones.

GRINDING WHEELS

The composition of a grinding wheel consists of the cutting material or abrasive (usually called "grit") and the bond. The cutting quality of a wheel depends chiefly on the grit and the hardness of the bonding material. The object of the bond is not only to hold the particles of grit together with the proper factor of safety but also to vary its tensile strength. The grinding wheel is called *hard* or *soft* depending on the tenacity with which the bond holds the particles together. A grinding wheel is said to be too hard when the bond retains the surface or cutting particles until they become dull, and it is said to be too soft when the particles are not held long enough to prevent undue wear of the wheel. Wheels are bonded by the vitrified, silicate, elastic, and rubber processes. The size of the grinder, such as the one shown in Fig. 1, is commonly taken from the diameter of the abrasive wheel or wheels. For example, a grinder with a 7-inch-diameter wheel is called a "seven-inch grinder."

Abrasive wheels are normally graded with respect to their "grit capacity." A grinding wheel that is to be used for sharpening tools should preferably be an aluminum oxide wheel of approximately

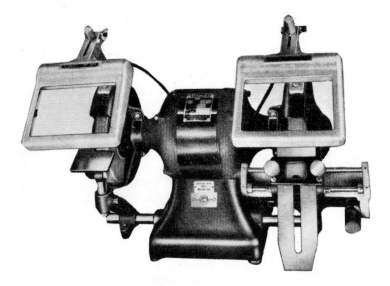

Fig. 1. A conventional tool grinder.

grade 60 grit and should be of a minimum hardness. A coarse, soft-grit wheel can remove material more rapidly than one with a finer grit, but the surface produced on the edge of the tool will be rougher than that produced with the fine-grit wheel. Therefore, to be suitable for grinding woodworking tools, a grinding wheel should be soft and should have a fine grit.

When using abrasive wheels for grinding woodworking tools, the high surface speed of the wheel in contact with the tool generates considerable heat; therefore, to reduce this heat production, the tool should be held lightly against the wheel and should frequently be dipped in water. If these two precautions are not taken, the edge of the tool may be severely burned, thereby reducing or even eliminating its usefulness. If a wheel that is running at a nominal speed has a tendency to burn the tool being sharpened, the wheel should be closely examined to determine whether it needs dressing. The power required to operate a 6- or 7-inch grinder is approximately ⅓ horsepower. On a direct-driven unit, the motor must run at approximately 3400 rpm in order to give the grinding wheel an efficient rim speed. For best results, the grinding wheel should run at a surface speed of from 5000 to 5500 fpm (feet per minute), but for grinding tools or when running the wheel in water, a slower speed is advisable. Use the following formula to calculate the rim speed of a grinding wheel:

$$S = \frac{M \times \pi \times D}{12}$$

where,

S=rim speed in feet per minute,
M=motor speed in revolutions per minute,
π=a constant—3.14159,
D=the diameter of the wheel in inches.

For an example of the usefulness of this formula, what is the surface (rim) speed of a 6-inch grinding wheel when it is directly driven by a 3400-rpm motor?

Using the formula above, we obtain

$$S = \frac{3400 \times 3.14159 \times 6}{12} = 5340 \ fpm$$

An essential feature of all grinding wheels is the wheel guards. These should enclose the wheel as completely as possible in order to prevent abrasive chips or larger fragments of the wheel from being thrown at the operator. Extra eye protection is also highly desirable when rough grinding and dressing operations are performed.

OILSTONES

These are used after the grinding operation to give the tool the highly keen edge necessary to cut wood smoothly. The oilstone is so called because oil is used on a whetting stone to carry off the heat resulting from friction between the stone and the tool and to wash away the particles of stone and steel that are worn off in the sharpening process. The process of rubbing the tool on

221

the stone is called "honing." There are two general classes of oilstones:

1. Natural.
2. Artificial.

Natural Oilstones

There are two general classes of natural stones grouped according to the locality in which they are found—Washita and Arkansas.

Ouachita Oilstone—*Ouachita* stone is found in the Ozark Mountains of Arkansas and is composed of almost pure silica. It is known throughout the world as the best natural stone for sharpening carpenters' and general woodworkers' tools. Its excellent sharpening qualities are due to small, sharp-pointed grains, or crystals, which are hexagonal in shape and are much harder than steel. *Ouachita* stone is found in various grades, from perfectly crystallized and porous grit to vitreous flint and hard sandstone. The sharpness of the grit depends entirely on its amount of crystallization. The best oilstones are made from extremely porous crystals.

Lily White Ouachita is the best selection or grading of natural Ouachita; it is perfectly white in color, uniform in texture, and nicely finished. *Rosy Red Ouachita* has an even, porous grit somewhat coarser than the *Lily White* grading and is therefore faster cutting. No. 1 *Ouachita* is a good oilstone for general use, where a medium-priced stone is desired. It is far superior to the many cheap, so-called "oilstones" on the market that are merely sandstones with a polished face; No. 1 *Ouachita* is not as uniform as the *Lily White* variety.

Arkansas Oilstone—Arkansas stone is composed of pure silica crystals, microscopic in size, and silica is among the hardest of known minerals. So hard and perfectly crystallized is the Arkansas stone that it is nearly sixteen times harder to cut than marble. The hardest of steel tools with the finest points or blades may be sharpened on the Arkansas stone without grooving the stone. Arkansas stone is prepared for commercial purposes in two grades, hard and soft.

Hard Arkansas is much harder than steel and will, therefore, cut away and sharpen steel tools. The extreme fineness of texture makes it a slow cutter but a perfect sharpener. *Soft Arkansas* is not quite as fine-grained and hard as the *Hard Arkansas,* but it cuts faster and is better for use by carvers, file makers, patternmakers, and all workers of hardwood.

Artificial Oilstones

These are made of carborundum (silicon carbide), emery, corundum (aluminum oxide), and other artificial abrasives, and are largely used in place of natural stones because they cut faster and may be made in any degree of fineness with an even texture.

Carborundum Oilstones—These stones are made from carborundum and may be used dry or with water or oil. They are quite porous and may be tempered clean and bright; they never fill or glaze. Corborundum stones are made in three grades, as follows:

Fine (FF)—For procuring a smooth, keen edge on tools made of hard steel.

Medium (180)—For sharpening tools quickly, where an extremely keen edge is not necessary.

Coarse (120)—To sharpen rather dull and large tools, which may later be finished with a fine stone, or in cases where a fine finish is not required.

India Oilstones—These are made from alundum. They possess the characteristics of hardness, sharpness, and toughness, as well as uniformity. They cut rapidly and are especially adaptable to the quick sharpening of all kinds of modern steel tools. All India stones are oil-filled by a patented process. This feature insures a moist, oily sharpening surface with the use of only a small quantity of oil. It also insures a good cutting surface by preventing the stone from filling with particles of steel. India stones are made in three grades, or grits, as follows:

Coarse—For sharpening large and rather dull or nicked tools, machine knives, and for general use where fast cutting is required without regard to fine finish.

Medium—For ordinary sharpening of tools that do not require a finished edge. Especially recommended for tools used in working softwoods, leather, and rubber.

Fine—For machinists and engravers, die workers, instrument workers, cabinetmakers, and all users of tools requiring an extremely fine, keen edge.

223

SUMMARY

Practically everyone knows the added pleasure that working with properly sharpened tools gives, but many try to maintain the edge of their tools by whetting when the original shape of the cutting edge is worn out or nicked and requires regrinding.

The grinder consists essentially of a horizontal spindle, the ends of which are threaded and fitted with flanges to take the grinding wheel. The spindle is either direct driven or has a conventional belt drive. The composition of a grinding wheel consists of the cutting material or abrasive, and the bond. The cutting quality of the grinding wheel depends on the grit and the hardness of the bonding material.

Oilstones are used after the grinding operation to give the tool the very keen edge necessary to cut wood smoothly. The oilstone is so called because oil is used to carry off the heat resulting from friction between the stone and the tool. There are natural stones and artificial stones, grouped according to the locality in which they are found.

REVIEW QUESTIONS

1. What are the two operations involved in sharpening woodworking tools?
2. Why are oilstones used in tool sharpening?
3. In what part of the country are most oilstones found?
4. What type of oilstone is best for whetting?

CHAPTER 19

How to Sharpen Tools

It cannot be stressed too much that edged tools must always be kept in perfect condition in order to do satisfactory work. This means that the cutting edge must be keen, free from nicks, and have the proper bevel. Sharpening is done by subjecting the tool to friction against an abrasive. The process includes grinding and honing.

GRINDING

First, the tool is placed on a grinding wheel in order to bring the bevel to the correct angle and to grind out any nicks that may exist in the cutting edge. Although this takes out the nicks and irregularities that are visible to the eye, the edge is still rough, as seen under a microscope. This roughness is considerably reduced by honing on an oilstone, although it is impossible to make the edge perfectly smooth because of the granular structure of the material.

When grinding tools on a stone without the use of either a guide or a rest, the tool is firmly pressed to and held at an angle of approximately 60° on the face of the revolving stone with both hands. Do not apply too much pressure, especially when grinding with the rapidly revolving emery wheels, since the operator is likely to burn the temper out of the tool. The edge of the tool must be continuously watched, especially with dry wheels. In case of a dry wheel, the tool must be immersed frequently in water to prevent overheating.

Plane-iron cutters vary in their make, temper, quality of steel, and uses, and they must be ground and sharpened for the sort of work which they are intended to execute. As previously explained, it is usual to grind a jack plane iron slightly curved, a fore plane iron almost flat, and a jointer or smoothing plane iron flat, except at the corners (see Fig. 15 of Chapter 15).

Before condemning any plane iron, therefore, carefully measure and compare the bevel of the cut and the thickness of the cutter. If the bevel is too long, the plane will jump and chatter. If it is too short, it will not cut, so it must be ground to a proper bevel. The length of the bevel should be twice the thickness of the iron (see Fig. 20 of Chapter 15).

Hatchets, axes, and adzes are always ground to their proper bevels; some have double and others have single bevels (see Fig. 4 of Chapter 14). When grinding, the blade is held to the stone surface with the right hand, and the handle is held with the left hand and on the left side, reversing the tool as the opposite side is being ground or sharpened.

Draw knives and spokeshave cutters are held with both hands, and the blade is kept horizontally flat on the stone as it is revolved toward the operator. Some woodworkers prefer to grind with the stone rotating toward the cutting edge, while others prefer to grind with the stone rotating away from the cutting edge. The latter is the safer method, because with the stone advancing, there is a danger of injury to the operator in case the tool digs into the stone. Do not use too much pressure with an advancing stone.

When sharpening tools and it is desired to have a keen edge, the tool should be honed after grinding on an oilstone.

HONING

After a tool has been ground on a grinding wheel, it will be found to have a wire edge. This edge must be removed, and the cutting edge must be made smooth by honing on an oilstone (see Fig. 21 of Chapter 15). The oilstone is constantly needed during all operations in carpentry in which the plane and chisel are used. It is needed more frequently than the grinding wheel, because the grinding wheel is only necessary when the tool becomes nicked, or when the edge becomes too dull to be sharpened on the oilstone without an undue amount of labor. The size of a typical oilstone for general use is approximately 2″ × 8″ or 2″ × 9″.

One rather desirable stone is the double carborundum, that is, a carborundum oilstone with one side coarse and the other side fine. Begin to hone on the coarse side and finish on the fine side with this type of stone. It is absolutely necessary to keep the oilstone clean and in perfect condition. If no attention is paid to this advice, experience will soon compel the amateur to do as directed. Oilstones should always be kept in their case when not in use. Use only a thin, clear oil on oilstones, and wipe the stone clean after using. Then moisten the stone with clean oil.

To clean an oilstone, wash it in kerosene; this will remove the gummed surface oil. This may be more easily and thoroughly done by heating the oilstone on a hot plate. A natural stone may also be heated on a hot plate to remove the surplus or gummed oil, after which a good cleaning with gasolene or ammonia will usually restore its cutting qualities; if this treatment does not work, scour the stone with a piece of loose emery or sandpaper which has been fastened to a perfectly smooth board.

When applying chisels and plane irons to an oilstone, the tool is held face up with both hands, the left hand in front, palm up, with the thumb on top, the fingers grasping the tool from underneath. The right hand is held behind the left hand, palm down, with the thumb under and the fingers reaching across the face of the tool. The blade edge is then rubbed back and forth with a sliding rotary motion on the face of the stone (which is first lubricated with oil or water). The rubbing angle is generally approximately 60°. After ten or twelve rubs, the blade is turned over and rubbed flat on the face side. The blade is then stropped; this may be done either by a slapping action or rubbed on a piece of old belting or leather set on top of the oilstone case. When this is done, the keenness of the blade may be tested with the thumb or by drawing the edge across the thumbnail, but this test must be done carefully to avoid injury.

Outside gouges are sharpened in the same manner as chisels. The tool should be rolled forward and backward when grinding the bevel. A whetstone is used to remove the wire edge by rubbing on the inside concave surface; the curved edge of the whetstone must fit exactly the arc of each gouge as closely as possible. Inside gouges must be ground on a curved stone and whetted to keen edges with the oilstones and whetstones.

Hollows and rounds, beading, and other special plane cutters are usually sharpened with whetstones and rarely require grinding. If they become nicked or injured on their edges, they are utterly useless.

Cold chisels, punches, and nail sets are best sharpened or pointed on grinding wheels. Carving tools are sharpened with small, fine whetstones.

When honing or whetting fine bench chisels, the burnished-face side must be kept perfectly flat on the face of the oilstone by pressing firmly down with the fingers of the left hand; the handle is held in the right hand. The rubbing action must be gentle and must also be rapidly repeated, turning the tool over constantly.

The edge of the chisel blade may slope slightly to the side of the oilstone, and it should be moved back and forth in a rotary motion on the stone. Do not raise the angle of the chisel too high on the stone, or the chisel will dig into and damage the surface of the oilstone. The oilstone should be wiped clean and reoiled frequently if several tools are to be sharpened.

SUMMARY

It is always important to keep a good working edge on woodworking tools in order to do satisfactory work. The cutting edge must always be free from nicks and have the proper bevel. Sharpening is done by subjecting the tools to grinding and honing.

The tools are placed on a grinding wheel in order to bring the bevel to the correct angle and to grind out the nicks and irregularities. After the grinding process the tool is then honed on an oilstone to remove the roughness.

REVIEW QUESTIONS

1. What angle is required on most cutting blades?
2. Explain the process of grinding the tools.
3. How do you clean an oilstone?
4. What is honing?

CHAPTER 20

How to Use the Steel Square

On most construction work, especially in house framing, the so-called "steel square" is invaluable for accurate measuring and for determining angles. The proper name is *framing square,* because the square with its markings was designed especially for marking timber in framing. However, the wrong name has become so firmly rooted that it will have to be put up with.

The square with its various scales and tables has been explained in Chapter 7. The function of this chapter is to explain these markings in more detail and also to explain their application by examples showing actual uses of the square. The following names are commonly used to identify the different portions of the square and should be noted and remembered:

Body—The long, wide member.
Face—The sides visible (both body and tongue) when the square is held by the tongue in the right hand with the body pointing to the left (see Fig. 1).
Tongue—The short, narrow member.
Back—The sides visible (both body and tongue) when the square is held by the tongue in the left hand with the body pointing to the right (see Fig. 1).

The square most generally employed has an 18-inch tongue and a 24-inch body. The body is 2 inches wide, and the tongue is $1\frac{1}{2}$ inches wide, $\frac{3}{16}$ inch thick at the heel or corner for strength, diminishing, for lightness, to the two extremities to approximately $\frac{3}{32}$ inch. The various markings on squares are of two kinds:

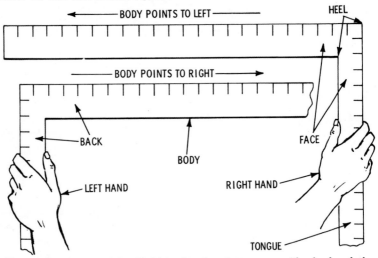

Fig. 1. The face and back sides of a framing square. The body of the square is sometimes called the blade.

1. Scales, or graduations.
2. Tables.

When buying a square, it is advisable to get one with all the markings rather than a cheap square on which the manufacturer has omitted some of the scales and tables. The following comparison illustrates the difference between a cheap and a complete square.

The square with the complete markings will cost more, but in the purchase of tools, you should make it a rule to purchase only the finest made. The general arrangements of the markings on squares differs somewhat with different makes; it is advisable to examine the different makes before purchasing to select the one best suited to your specific requirements.

APPLICATION OF THE SQUARE

As stated previously, the markings on squares of different makes sometimes vary both in their position on the square and the mode of application. However, a thorough understanding of the application of the markings on any first-class square will enable the student to easily acquire proficiency with any other square.

	Tables	Graduations
Cheap Square	Rafter, Essex, Brace	1/16, 1/12, 1/8, 1/4
Complete Markings	Rafter, Essex, Brace, Octagon, Polygon cuts	1/100, 1/64, 1/32, 1/16, 1/12, 1/10, 1/8, 1/4

Scale Problems

The term "scales" is used to denote the inch divisions of the tongue and body length found on the outer and inner edges; the inch graduations are divided into $\frac{1}{4}$, $\frac{1}{8}$, $\frac{1}{10}$, $\frac{1}{12}$, $\frac{1}{16}$, $\frac{1}{32}$, $\frac{1}{64}$, and $\frac{1}{100}$. All these graduations should be found on a first-class square. The various scales start from the heel of the square, that is, at the intersection of the two outer, or two inner, edges.

A square with only the scale markings is adequate to solve many problems that arise when laying out carpentry work. An idea of its range of usefulness is shown in the following problems.

Problem 1—To describe a semicircle given the diameter.

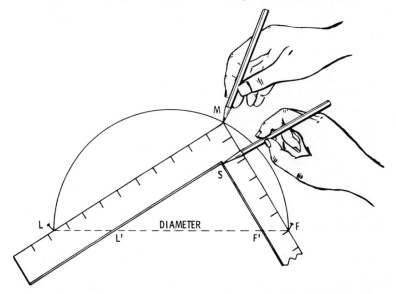

Fig. 2. Problem 1. The outer-heel method is described in the text. For the inner-heel method, the pencil is held at S, and the distance L'F' should be taken to equal the diameter, with the inner edges of the square sliding on the brads.

Drive brads at the ends of the diameter LF, as shown in Fig. 2. Place the outer edges of the square against the nails, and hold a lead pencil at the outer heel M; any semicircle can then be described, as indicated. This is the outer-heel method, but a better guide for the pencil is obtained by using the inner-heel method, which is also shown in the figure.

Problem 2—To find the center of a circle.

Lay the square on the circle so that its outer heel lies in the circumference. Mark the intersections of the body and tongue with the circumference. The line that connects these two points is a diameter. Draw another diameter (obtained in the same way); the intersection of the two diameters is the center of the circle, as shown in Fig. 3.

Problem 3—To describe a circle through three points which are not in a straight line.

Join the three points with straight lines; bisect these lines, and, at the points of bisection, erect perpendiculars with the square. The intersection of these perpendiculars is the center from which a circle may be described through the three points, as shown in Fig. 4.

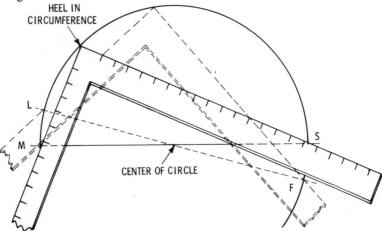

HEEL IN
CIRCUMFERENCE

L

M

S

CENTER OF CIRCLE

F

Fig. 3. Problem 2. Draw diameters through points LF and MS, where the sides of the square touch the circle with the heel in the circumference. The intersection of these two lines is the center of the circle.

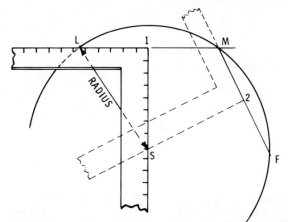

Fig. 4. Problem 3. Let points L, M, and F be three points which are not in a straight line. Draw lines LM and MF, and bisect them at points 1 and 2, respectively. Apply the square with the heel at points 1 and 2, as shown; the intersection of the perpendicular lines thus obtained, point S, is the center of the circle. Lines LS, MS, and FS represent the radius of the circle, which may now be described through points L, M, and F.

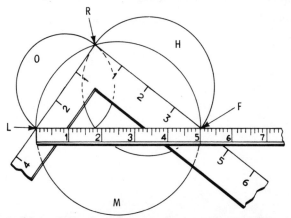

Fig. 5. Problem 4. Let O and H be the two given circles, with their diameters LR and RF at right angles. Suppose the diameter of O is 3 inches and the diameter of H is 4 inches. Points L and F, at these distances from the heel of the square, will be 5 inches apart, as measured with a 2-foot rule. This distance LF, or 5 inches, is the diameter of the required circle. Proof: $(LF)^2 = (LR)^2 + (RF)^2$, or $25 = 9 + 16$.

Problem 4—To find the diameter of a circle whose area is equal to the sum of the areas of two given circles.

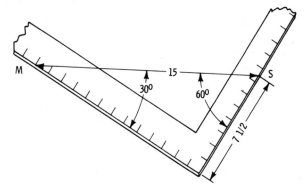

Fig. 6. Problem 5. Draw line MS, 15 inches long. Place the square so that point S touches the tongue 7-1/2 inches from the heel and point M touches the body. The triangle thus formed will have an angle of 30° at M and an angle of 60° at S.

Lay off on the tongue of the square the diameter of one of the given circles, and on the body the diameter of the other circle. The distance between these points (measure across with a 2-foot rule) will be the diameter of the required circle, as shown in Fig. 5.

Problem 5—To lay off angles of 30° and 60°.

Mark off 15 inches on a straight line, and lay the square so that the body touches one end of the line and the 7½-inch mark on

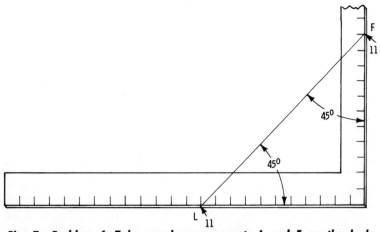

Fig. 7. Problem 6. Take equal measurements L and F on the body and tongue of the square. The triangle thus formed will have an angle of 45° at L and at F.

the tongue is against the other end of the line, as shown in Fig. 6. The tongue will then form an angle of 60° with the line, and the body will form an angle of 30° with the line.

Problem 6—To lay off an angle of 45°.

The diagonal line connecting equal measurements on either arm of the square forms angles of 45° with the blade and tongue, as shown in Fig. 7.

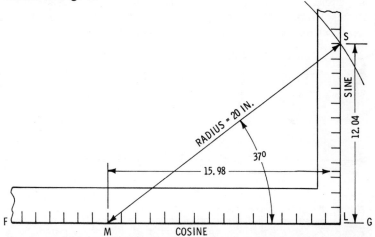

Fig. 8. Problem 7. Let 37° be the required angle. Place the body of the square on line FG, and, from Table 1, lay off LS (12.04) on the tongue and LM (15.98) on the body. Draw line MS; then angle LMS = 37°. Line MS will be found to be equal to 20 inches for any angle, because the values given in Table 1 for LS and MS are natural sines and cosines multiplied by 20.

Problem 7—To lay off any angle.

Table 1 gives the values for measurements on the tongue and the body of the square so that by joining the points corresponding to the measurements, any angle may be laid out from 1° to 45°, as explained in Fig. 8.

Problem 8—To find the octagon of any size timber.

Place the body of a 24-inch square diagonally across the timber so that both extremities (ends) of the body touch opposite edges. Make a mark at 7 inches and 17 inches, as shown in Fig. 9. Repeat this process at the other end, and draw lines through the pairs of marks. These lines show the portion of material that must be taken off the corners.

The side of an inscribed octagon can be obtained from the side of a given square by multiplying the side of the square by 5 and dividing the product by 12. The quotient will be the side of the octagon. This method is illustrated in Fig. 10.

The side of a hexagon is equal to the radius of the circumscribing circle. If the side of a desired hexagon is given, arcs should be struck from each extremity at a radius equal to its length. The point where these arcs intersect is the center of the circumscribing

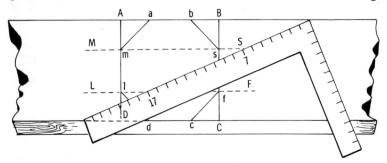

Fig. 9. Problem 8. Lay out square ABCD. Place the body of a 24-inch square as shown, and draw parallel lines MS and LF through points 7 and 17. These lines intercept sides ml and sf of the octagon. To lay off side sb, place the square so that the tongue touches point s and the body touches l, with the heel touching line AB. The remaining sides are obtained in a similar manner.

circle, and having described it, it is sufficient to lay off chords on its circumference equal to the given side to complete the hexagon.

Square-and-Bevel Problems

By the application of a large bevel to the framing square, the combined tool becomes a calculating machine, and by its use, arithmetical processes are greatly simplified. The bevel is preferably made of steel blades. The following points should be observed in its construction:

The edges of each blade must be true; blade E in Fig. 11 must lie under the square so that it does not hide the graduations; the two blades must be fastened by a thumbscrew to lock them together; blade L should have a hole near each end and one in

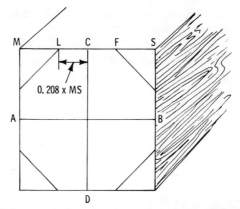

Fig. 10. Problem 8 (second method). Let lines AB and CD be center lines, and let line MS be one side of the square timber. Multiply the length of the side by 0.208; the product is half the side of the inscribed octagon. Therefore, lay off CF and CL, each 0.208 times side MS; LF is then one side of the octagon. Set dividers to distance CL, and lay off the other sides of the octagon from the center lines to complete the octagon.

Table 1.   Angle Table for the Square

Angle	Tongue	Body	Angle	Tongue	Body	Angle	Tongue	Body
1	0.35	20.00	16	5.51	19.23	31	10.28	17.14
2	0.70	19.99	17	5.85	19.13	32	10.60	16.96
3	1.05	19.97	18	6.18	19.02	33	10.89	16.77
4	1.40	19.95	19	6.51	18.91	34	11.18	16.58
5	1.74	19.92	20	6.84	18.79	35	11.47	16.38
6	2.09	19.89	21	7.17	18.67	36	11.76	16.18
7	2.44	19.85	22	7.49	18.54	37	12.04	15.98
8	2.78	19.81	23	7.80	18.40	38	12.31	15.76
9	3.13	19.75	24	8.13	18.27	39	12.59	15.54
10	3.47	19.70	25	8.45	18.13	40	12.87	15.32
11	3.82	19.63	26	8.77	17.98	41	13.12	15.09
12	4.16	19.56	27	9.08	17.82	42	13.38	14.89
13	4.50	19.49	28	9.39	17.66	43	13.64	14.63
14	4.84	19.41	29	9.70	17.49	44	13.89	14.39
15	5.18	19.32	30	10.00	17.32	45	14.14	14.14

the middle, so that blade E may be shifted as required, with a large notch near each hole in order to observe the position of blade E.

Problem 9—To find the diagonal of a square.

237

Set blade E to 10⅜ on the tongue and to 15 on the body. Assume an 8-inch square. Slide the bevel sideways along the tongue until blade E is against point 8. The other edge will touch 11⁵⁄₁₆ on the body; this is the required diagonal.

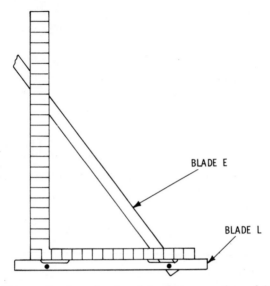

Fig. 11. The application of a bevel to the square for solving square-and-bevel problems.

Problem 10—To find the circumference of a circle from its diameter.

Set the bevel blade to 7 on the tongue of the square and to 22 on the body. The reading on the body will be the circumference corresponding to the diameter at which E is set on the tongue. To reverse the process, use the same bevel, and read the required diameter from the tongue, the circumference being set on the body.

Problem 11—Given the diameter of a circle, find the side of a square of equal area.

Set the bevel blade to 10⅝ on the tongue and to 12 on the body. The diameter of the circle, on the body, will give the side of the equal square on the tongue. If the circumference is given instead of the diameter, set the bevel to 5½ on the tongue and to 19½ on the body, thereby finding the side of the square on the tongue.

238

Problem 12—Given the side of a square, find the diameter of a circle of equal area.

Using the same bevel as in Problem 11, blade E is set to the given side on the tongue of the square, and the required diameter is read off the body.

Problem 13—Given the diameter of the pitch circle of a gear wheel and the number of teeth, find the pitch.

Take the number of teeth, or a proportional part, on the body of the square and the diameter, or a similar proportional part, on the tongue, and set the bevel blade to those marks. Slide the bevel to 3.14 on the body, and the number given on the tongue multiplied by the proportional divisor will be the required pitch.

Problem 14—Given the pitch of the teeth and the diameter of the pitch circle in a gear wheel, find the number of teeth.

Set the bevel blade to the pitch on the tongue and to 3.14 on the body of the square. Move the bevel until it marks the diameter on the tongue. The number of teeth can then be read from the blade. If the diameter is too large for the tongue, divide it and the pitch into proportional parts, and multiply the number found by the same figure.

Problem 15—The side of a polygon being given, find the radius of the circumscribing circle.

Table 2.   Inscribed Polygons

Number of Sides	3	4	5	6	7	8	9	10	11	12
Radius	56	70	74	60	60	98	22	89	80	85
Side	97	99	87	60	52	75	15	95	45	44

Set the bevel to the pairs of numbers in Table 2, taking one-eighth or one-tenth of an inch as a unit. The bevel, when locked, is slid to the given length of the side, and the required length of the radius is read on the other leg of the square. For example, if a pentagon (5 sides) must be layed out with a side of 6 inches, the bevel is set to the figures in column 5 with the lesser number set on the tongue. In this case, $7\frac{4}{8} = 9\frac{1}{4}$ on the tongue, and $8\frac{7}{8} = 10\frac{7}{8}$ on the body of the square. Slide the bevel to 6 on the body. The length of the radius, $5\frac{3}{32}$, will be read on the tongue.

Problem 16—To divide the circumference of a circle into a given number of equal parts.

Table 3. Chords or Equal Parts

No. of Parts		Y	Z	No. of Parts	Y	Z	No. of Parts	Y	Z
3	Triangle	1.732	.5773	15	.4158	2.4050	40	.1569	6.3728
4	Square	1.414	.7071	16	.3902	2.5628	45	.1395	7.1678
5	Pentagon	1.175	.8006	17	.3675	2.7210	50	.1256	7.9618
6	Hexagon	1.000	1.0000	18	.3473	2.8793	54	.1163	8.5984
7	Heptagon	.8677	1.1520	19	.3292	3.0376	60	.1047	9.5530
8	Octagon	.7653	1.3065	20	.3129	3.1962	72	.0872	11.462
9	Nonagon	.6840	1.4619	22	.2846	3.5137	80	.0785	12.733
10	Decagon	.6180	1.6184	24	.2610	3.8307	90	.0698	14.327
11	Undecagon	.5634	1.7747	25	.2506	3.9904	100	.0628	15.923
12	Duodecagon	.5176	1.9319	27	.2322	4.3066	108	.0582	17.182
13	Tridecagon	.4782	2.0911	30	.2090	4.7834	120	.0523	19.101
14	Tetradecagon	.4451	2.2242	36	.1743	5.7368	150	.0419	23.866

From the column marked Y in Table 3, take the number opposite the given number of parts. Multiply this number by the radius of the circle. The product will be the length of the cord to lay off on the circumference.

Problem 17—Given the length of a cord, find the radius of the circle.

This is the same as Problem 16, but the present form may be more expeditious for calculations. The method is useful for determining the diameter of gear wheels when the pitch and number of teeth have been given. Multiply the length of the cord, width of the side, or pitch of the tooth by the figures found corresponding to the number of parts in column Z of Table 3. The result is the radius of the desired circle.

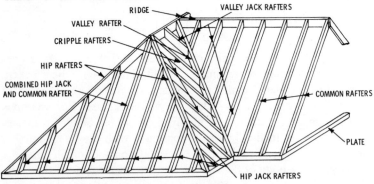

Fig. 12. A typical roof frame, showing the ridge, the plate, and various types of rafters.

240

TABLE PROBLEMS

The term "table" is used here to denote the various markings on the framing square with the exception of the scales already described. Since these tables relate mostly to problems encountered in cutting lumber for roof-frame work, it is first necessary to know something about roof construction so as to be familiar with the names of the various rafters and other parts. Fig. 12 is a view of a roof frame showing the various members. In the figure it will be noted that there is a plate at the bottom and a ridge timber at the top; these are the main members to which the rafters are fastened.

Main or Common Rafters

The following definitions relating to rafters should be carefully noted:

The **rise** of a roof is the distance found by following a plumb line from a point on the central line of the top of the ridge to the level of the top of the plate.

The **run** of a common rafter is the shortest horizontal distance from a plumb line through the center of the ridge to the outer edge of the plate.

The **rise per foot run** is the basis on which rafter tables on some squares are made. The term is self-defining. Other roof components are illustrated in Fig. 13.

To obtain the rise per foot run, multiply the rise by 12 and divide by the run; thus:

$$rise \ per \ foot \ run = \frac{rise \times 12}{run}$$

The factor 12 is used to obtain a value in inches, since the rise and run are normally given in feet.

Example—If the rise is 8 feet and the run is 8 feet, what is the rise per foot run?

$$rise \ per \ foot \ run = \frac{8 \times 12}{8} = 12 \ inches$$

The rise per foot run is always the same for a given pitch and can be readily remembered for all ordinary pitches; thus:

Pitch	½	⅓	¼	⅙
Rise per foot run (in.) ...	*12*	*8*	*6*	*4*

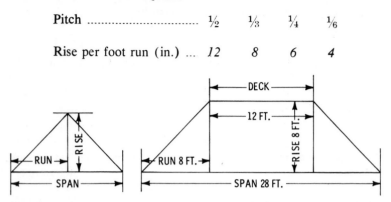

Fig. 13. The terms rise, run, span, and deck are illustrated in two types of roofs. If the rafters rise to a deck instead of a ridge, subtract the width of the deck from the span. For example, assume the span is 28 feet and the deck is 12 feet; the difference is 16 feet, and the pitch = 8/(28 − 12) = 1/2.

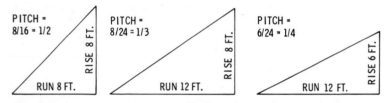

Fig. 14. To obtain the pitch of any roof divide the rise of the rafters

by twice the run.

The pitch can be obtained if the rise and run are known, as shown in Fig. 14, by dividing the rise by twice the run, or

$$pitch = \frac{rise}{2 \times run}$$

In roof construction, the rafter ends are cut with slants which rest against the ridge and the plate, as shown in Fig. 15A. The slanting cut which rests against the ridge board is called the *plumb,* or *top,* cut, and the cut which rests on the plate is called the *seat,* or *heel,* cut.

The length of the common rafter is the length of a line from the outer edge of the plate to the top corner of the ridge board or, if there is no ridge board, from the outer edge of the plate to the vertical center line of the building, as shown in Fig. 15B. The run of the rafter, then, in the first case is one-half the width of the building less one-half the thickness of the ridge, if any; if there is no ridge board, the run is one-half the width of the building. Where there is a deck, the run of the rafters is one-half the width of the building less one-half the width of the deck.

Now, with a 24-inch square, draw diagonals connecting 12 on the tongue (corresponding to the run) to the value from Table 4 on the body (corresponding to the rise) to obtain the

Table 4. Pitch Table

Pitch	1	11/12	5/6	3/4	2/3	7/12	1/2	5/12	1/3	1/4	1/6	1/12
Run	12	12	12	12	12	12	12	12	12	12	12	12
Rise	24	22	20	18	16	14	12	10	8	6	4	2

pitch angle for any combination of run and rise. This procedure is further illustrated in Fig. 16.

Hip Rafters

The hip rafter represents the hypotenuse, or diagonal, of a right-angle triangle; one side is the common rafter, and the other side is the plate, or that part of the plate lying between the foot of the hip rafter and the foot of the adjoining common rafter, as shown in Fig. 17.

The rise of the hip rafter is the same as that of the common rafter. The run of the hip rafter is the horizontal distance from the plumb line of its rise to the outside of the plate at the foot of the hip rafter. This run of the hip rafter is to the run of the common rafter as 17 is to 12. Therefore, for a ⅙ pitch, the common rafter run and rise are 12 and 4, respectively, while the hip rafter run and rise are 17 and 4, respectively.

For the top and bottom cuts of the common rafter, the figures are used that represent the common rafter run and rise, that is, 12 and 4 for a ⅙ pitch, 12 and 6 for a ¼ pitch, etc. However, for the top and bottom cuts of the hip rafter, use the figures 17 and 4, 17 and 6, etc., as the run and rise of the hip rafter. It must be remembered, however, that these figures will not be correct if the pitches on the two sides of the hip (or valley) are not the same.

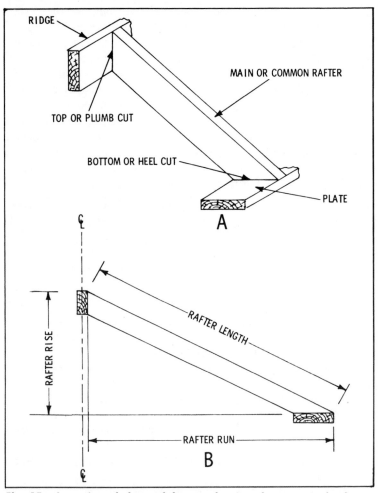

Fig. 15. A portion of the roof frame, showing the top, or plumb, cut and the bottom, or heel, cut is illustrated in A. The length of a common rafter is shown in B.

Valley Rafters

The valley rafter is the hypotenuse of a right-angle triangle made by the common rafter with the ridge, as shown in Fig. 18. This corresponds to the right-angle triangle made by the hip rafter with the common rafter and plate. Therefore, the rules for

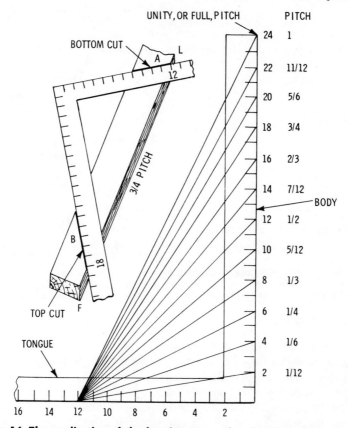

Fig. 16. The application of the framing square for obtaining the various pitches given in Table 4.

the lengths and cuts of valley rafters are the same as for hip rafters.

Jack Rafters

These are usually spaced either 16 or 24 inches apart, and, since they lie equally spaced against the hip or valley, the second jack rafter must be twice as long as the first, the third three times as long as the first, and so on, as shown in Fig. 19. One reason for the 16- and 24-inch spacings on jack rafters is because of the roof sheathing; therefore, the rafters must be 16 or 24 inches apart so that the sheathing may be conveniently nailed to it.

245

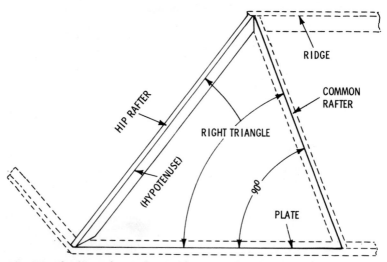

Fig. 17. The hip rafter is framed between the plate and the ridge and is the hypotenuse of a right-angle triangle whose other two sides are the adjacent common rafter and the intercepted portion of the plate.

Cripple Rafters

A cripple rafter is a jack rafter which touches neither the plate nor the ridge; it extends from the valley rafter to the hip rafters. The cripple-rafter length is that of the jack rafter plus the length necessary for its bottom cut, which is a plumb cut similar to the top cut. Top and bottom (plumb) cuts of cripples are the same as the top cut for jack rafters. The side cut at the hip and valley are the same as the side cut for jacks.

Finding Rafter Lengths Without the Aid of Tables

In the directions accompanying framing squares and in some books, frequent mention is made of the figures 12, 13, and 17. The reader is told for common rafters to "use figure 12 on the body and the rise of the roof on the tongue"; for hip or valley rafters, "use figure 17 on the body and the rise of the roof on the tongue"—and no explanation of how these fixed numbers are obtained is provided. The intelligent workman should not be satisfied with knowing which number to use, but he should want to know *why* each particular number is used. This can be readily understood by referring to Fig. 20. In this illustration, let ABCD

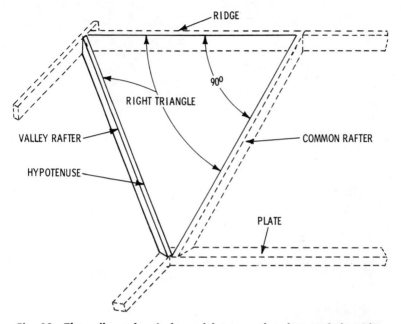

Fig. 18. The valley rafter is framed between the plate and the ridge and is the hypotenuse of a right-angle triangle whose other two sides are the adjacent common rafter and the intercepted portion of the

ridge board.

be a square whose sides are 24 inches long, and let abcdefgL be an inscribed octagon. Each side of the octagon (ab, bc, etc.) measures 10 inches; that is, LF = one-half side = 5 inches, and by construction, FM = 12 inches. Now, let FM represent the run of a common rafter. Then LM will be the run of an octagon rafter, and DM will be the run of a hip or valley rafter. The values for the run of octagon and hip or valley rafters (LM and DM, respectively) are obtained as follows:

$$LM = \sqrt{(FM)^2 + (LF)^2} = \sqrt{(12)^2 + (5)^2} = 13$$

$$DM = \sqrt{(FM)^2 + (DF)^2} = \sqrt{(12)^2 + (12)^2} = 16.97,$$
$$or \ approximately \ 17$$

247

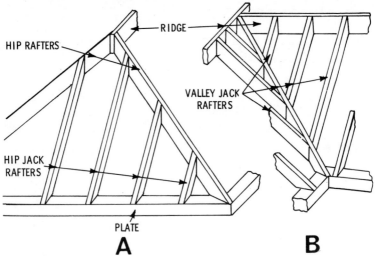

Fig. 19. Jack rafters. A, hip jack rafters, framed between the plate and hip rafters; B, valley jack rafters, framed between the ridge and the valley rafter.

Example—What is the length of a common rafter having a 10-foot run and a $\frac{3}{8}$ pitch?

For a 10-foot run,

$$\textit{the span} = 2 \times 10 = 20 \textit{ feet}$$

with $\frac{3}{8}$ pitch,

$$\textit{rise} = {}^3{}_8 \times 20 = 7.5 \textit{ feet}$$

$$\textit{rise per foot run} = \frac{\textit{rise} \times 12}{\textit{run}} = \frac{7.5 \times 12}{10} = 9 \textit{ inches}$$

On the body of the square shown in Fig. 21, take 12 inches for 1 foot of run, and on the tongue, take 9 inches for the rise per foot of run. The diagonal, or distance between the points thus obtained, will be the length of the common rafter per foot of run with a $\frac{3}{8}$ pitch. The distance FM measures 15 inches, or by calculation;

$$FM = \sqrt{(12)^2 + (9)^2} = 15 \textit{ inches}$$

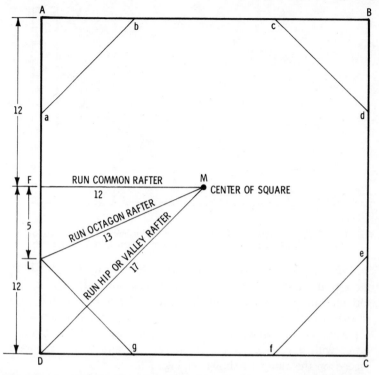

Fig. 20. A square and an inscribed octagon are used to illustrate the method of obtaining and using points 12, 13, and 17 in the application of a framing square to determine the length of rafters without the aid of rafter tables

Since the length of run is 10 feet,

$$length\ of\ rafter = length\ of\ run \times length\ per\ foot$$

$$= 10 \times {}^{15}\!/_{12}$$

$$= \frac{150}{12}$$

$$= 12.5\ feet$$

The combination of figures 12 and 9 on the square, as shown in Fig. 21, not only gives the length of the rafter per foot of run,

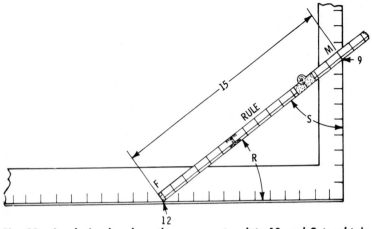

Fig. 21. A rule is placed on the square at points 12 and 9 to obtain the length of a common rafter per foot of run with a 3/8 pitch.

but, if the rule is considered as the rafter, the angles S and R for the top and bottom cuts are obtained. The points for making the top and bottom cuts are found by placing the square on the rafter so that a portion of one arm of the square represents the run and a portion of the other arm represents the rise. For the common rafter with a ⅜ pitch, these points are 12 and 9; the square is placed on the rafter as shown in Fig. 22.

Example—What length must an octagon rafter be to join a common rafter having a 10-foot run (as rafters MF and ML in Fig. 20)?

From Fig. 20, it is seen that the run per foot of an octagon rafter, as compared with a common rafter, is as 13 is to 12, and that the rise for a 13-inch run of an octagon rafter is the same as for the run of a 12-inch common rafter. Therefore, measure across from points 13 and 9 on the square, as MS in Fig. 23, which gives the length (15¾ inches) of an octagon rafter per foot of run of a common rafter. The length multiplied by the run of a common rafter gives the length of an octagon rafter; thus:

$$15\tfrac{3}{4} \times 10 = 157\tfrac{1}{2} \ inches = 13 \ feet, \ 1\tfrac{1}{2} \ inches$$

Points 13 and 9 on the square (MS in Fig. 23) give the angles for the top and bottom cuts.

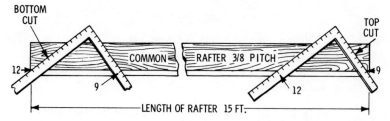

Fig. 22. *The square is placed on the rafter at points 12 and 9, as shown, thereby giving the proper angles for the bottom and top cuts.*

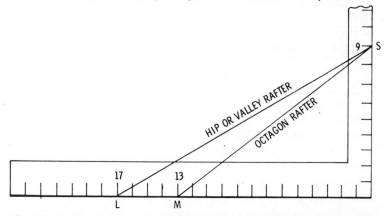

Fig. 23. *Measurements using the square for octagon and hip or valley rafters, illustrating the use of points 13 and 17. Line MS (13,9) is the octagon rafter length per foot of run of a common rafter with a 3/8 pitch; line LS (17,9) is the hip or valley rafter length per foot of run of a common rafter with a 3/8 pitch.*

Example—What length must a hip or valley rafter be to join a common rafter having a 10-foot run (as rafters MF and MD in Fig. 20)?

Fig. 20 shows that the run per foot of a hip or valley rafter, as compared with a common rafter, is as 17 is to 12, and that the rise per 17-inch run of a hip or valley rafter is the same as for a 12-inch run of a common rafter. Therefore, measure across from points 17 and 9 on the square, as LS in Fig. 23; this gives the length (19¼ inches) of the hip or valley rafter per foot of common rafter. This length, multiplied by the run of a common rafter gives the length of the hip or valley rafter; thus:

251

$$19\tfrac{1}{4} \times 10 = 192\tfrac{1}{2} \text{ inches} = 16 \text{ feet, } \tfrac{1}{2} \text{ inch}$$

Points 17 and 9 on the square (LS in Fig. 23) gives the angles for the top and bottom cuts.

Table 5 gives the points on the square of the top and bottom cuts of various rafters.

Table 5.   Square Points for Top and Bottom Cuts

PITCH	1	11/12	5/6	3/4	2/3	7/12	1/2	5/12	1/3	1/4	1/6	1/12
Tongue — Common							12					
Tongue — Octagon							13					
Tongue — Hip or Valley							17					
Body	24	22	20	18	16	14	12	10	8	6	4	2

RAFTER TABLES

The arrangement of these tables varies considerably with different makes of squares, not only in the way they are calculated but also in their positions on the square. On some squares, the rafter tables are found on the face of the body; on others, they are found on the back of the body. There are two general classes of rafter tables, grouped as follows:

1. Length of rafter per foot of run.
2. Total length of rafter.

Evidently, where the total length is given, there is no figuring to be done, but when the length is given per foot of run, the reading must be multiplied by the length of run to obtain the total length of the rafter. To illustrate these differences, directions for using several types of squares are given in the following sections. These differences relate to the common and hip or valley rafter tables.

Reading the Total Length of the Rafter

One popular type of square is selected as an example to show how rafter lengths may be read directly without any figuring. The rafter tables on this particular square occupy both sides of the body instead of being combined in one table; the common rafter table is found on the back, and the hip, valley, and jack rafter tables are located on the face.

Common Rafter Table—The common rafter table, Fig. 24, includes the outside-edge graduations of the back of the square on both the body and the tongue; these graduations are in twelfths. The inch marks may represent inches or feet, and the

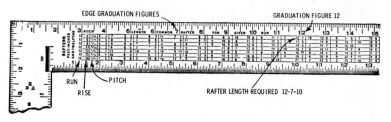

Fig. 24. The common rafter table.

twelfths marks may represent twelfths of an inch or twelfths of a foot (inches). The edge-graduation figures above the table represent the run of the rafter; under the proper figure on the line representing the pitch is found the rafter length required in the table. The pitch is represented by the figures at the left of the table under the word **PITCH**; thus:

12 Feet of Run						
Feet of Rise	4	6	8	10	12	15	18
Pitch	1/6	1/4	1/3	5/12	1/2	5/8	3/4

The length of a common rafter given in the common rafter table is from the top center of the ridge board to the outer edge of the plate. In actual practice, deduct one-half the thickness of the ridge board, and add for any eave projection beyond the plate.

Example—Find the length of a common rafter for a roof with a $\frac{1}{6}$ pitch (rise = $\frac{1}{6}$ the width of the building) and a run of 12 feet (found in the common rafter table, Fig. 24, the upper, or $\frac{1}{6}$-pitch ruling).

Find the rafter length required under the graduation figure 12. This is found to be 12, 7, 10, which means 12 feet, $7\frac{10}{12}$ inches. If the run is 11 feet and the pitch is $\frac{1}{2}$ (the rise = $\frac{1}{2}$ the width of the building), then the rafter length will be 15, 6, 8,

253

which means 15 feet $6\frac{8}{12}$ inches. If the run is 25 feet, add the rafter length for a run of 20 feet to the rafter length for a run of 5 feet. When the run is in inches, then in the rafter table read inches and twelfths instead of feet and inches. For instance, if, with a $\frac{1}{2}$ pitch, the run is 12 feet 4 inches, add the rafter length of 4 inches to that of 12 feet as follows:

For a run of 12 feet, the rafter length is 16 feet, $11\frac{8}{12}$ inches.
For a run of 4 inches, the rafter length is $\qquad$ $5\frac{8}{12}$ inches.

Total—17 feet, $5\frac{4}{12}$ inches.

The run of 4 inches is found under the graduation 4 and is 5, 7, 11, which is approximately $5\frac{8}{12}$ inches. If the run was 4 feet, it would be read as 5 feet, $7\frac{11}{12}$ inches.

Hip Rafter Table—This table, as shown in Fig. 25, is located on the face of the body and is used in the same manner as the

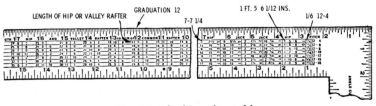

Fig. 25. The hip rafter table.

table for common rafters explained above. In the hip rafter table, the outside-edge-graduation figures represent the run of common rafters. The length of a rafter given in the table is from the top center of the ridge board to the outer edge of the plate. In actual practice, deduct one-half the thickness of the ridge board, and add for any eave projection beyond the plate. When using this table, find the figures on the line with the required pitch of the roof.

Under **PITCH,** the set of three columns of figures gives the pitch. The seven pitches in common use are given, as for example $\frac{1}{6}$-12-4; this means that for a $\frac{1}{6}$-pitch, there is a 12-inch run per 4-inch rise.

Under **HIP**, the set of figures gives the length of the hip and valley rafter per foot of run of common rafter for each pitch, as 1 foot, $5\frac{6}{12}$ inches for a $\frac{1}{6}$ pitch.

Under **JACK** (16 inches on center), the set of figures gives the length of the shortest jack rafter, spaced 16 inches on center, which is also the difference in length of succeeding jack rafters.

Example—If the jack rafters are spaced 16 inches on center for a $\frac{1}{6}$-pitch roof, find the lengths of the jacks and cut bevels.

The jack top and bottom cuts (or plumb and heel cuts) are the same as for the common rafter. Take 12 on the tongue of the square; that is, mark on the $9\frac{1}{2}$ sides, as shown in the illustration. which represents the rise per foot of the roof, or, if the pitch is given, take the figures in Table 5 that correspond to the given pitch. Thus, for a $\frac{1}{6}$ pitch, these points are 12 and 4. Fig. 26 shows the square on the jack in this position for marking top and bottom cuts.

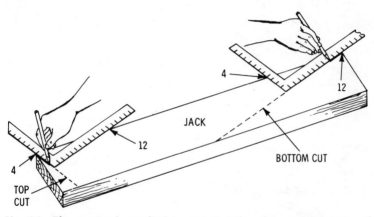

Fig. 26. The square is applied to a jack rafter for marking top and bottom cuts. The vertical and horizontal cuts for jack rafters are the same as for common rafters.

Look along the line of $\frac{1}{6}$ pitch, in Fig. 27, under **JACK** (16-inch center), and find $16\frac{7}{8}$, which is the length in inches of the shortest jack and is also the amount to be added for the second jack. Deduct one-half the thickness of the hip rafter, because the jack rafter lengths given in this table are to centers. Also, add for any projection beyond the outer edge of the plate.

Look along the line of $\frac{1}{6}$ pitch, in Fig. 27, under **JACK** (side cut), and find $9 - 9\frac{1}{2}$ for a $\frac{1}{6}$ pitch. These figures refer to the graduated scale on the edge of the arm of the square. To obtain

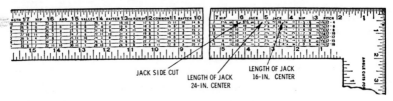

JACK SIDE CUT

LENGTH OF JACK
24-IN. CENTER

LENGTH OF JACK
16-IN. CENTER

Fig. 27. A rafter table.

the required bevel, take 9 on one arm and 9½ on the other, as shown in Fig. 28. It should be carefully noted that the last figure,

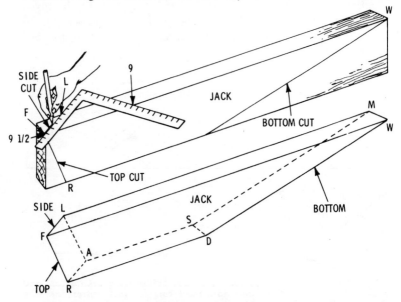

Fig. 28. Marking and cutting a jack rafter with the aid of the square. FR and DW are the marks for the top and bottom cuts, respectively. With the jack rafter cut as marked, LARF represents the section cut at the top, and MSDW represents the section cut at the bottom.

or figure to the right, gives the point on the marking side of the square; that is, mark on the 9½ sides, as shown in the illustration.

Under **JACK** (24 inches on center), the set of figures gives the length of the shortest jack rafter spaced 24 inches on center, which is also the difference in length of succeeding jack rafters. Deduct one-half the thickness of the hip or valley rafter, because the jack rafter lengths given in the table are to centers. Also, add for any projection beyond the plate.

Under **HIP**, the set of figures gives the side cut of the hip and valley rafters against the ridge board or deck, as 7–7¼ for a ⅙ pitch (mark on the 7¼ side).

To get the cut of the sheathing and shingles (whether hip or valley), reverse the figures under **HIP**, as 7¼–7 instead of 7–7¼. For the hip top and bottom cuts, take 17 on the body of the square, and, on the tongue, take the figure which represents the rise per foot of the roof.

Fig. 29 shows the marking and cut of the hip rafter, and Fig. 30 shows the rafter in position resting on the cap and the ridge. The section L′A′R′F′ resting on ridge is the same as L′A′R′F′ in Fig. 29.

Under **HIP AND VALLEY**, the set of figures gives the length of run of the hip or valley rafter for each pitch of the common rafter. For instance, for a roof with a ⅙ pitch under the figure 12

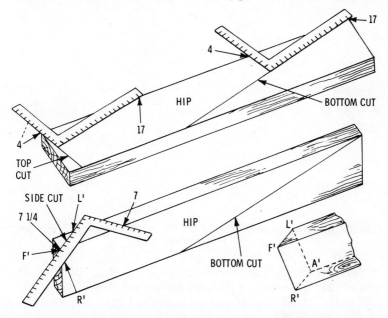

Fig. 29. The square, as applied to hip rafters, for marking top, bottom, and side cuts. Note that the number 17 on the body is used for hip rafters. Section L′A′R′F′ shows the bevel required for the ridge.

(representing the run of the common rafter, or one-half the width of the building), along the ⅙-pitch line of figures find 17, 5, 3, which means 17 feet, 5³⁄₁₂ inches, which is the length of the hip or valley rafter. Deduct one-half the thickness of the ridge board, and add for eave overhang beyond the plate, which is the length of the hip or valley rafter required for a roof with a ⅙ pitch and a common rafter run of 12 feet.

Example—Find the length of the hip rafter for a building that has a 24-foot span and a ⅙ pitch (a 4-inch rise per foot of run).

In the hip rafter table (Fig. 25) along the line of figures for ⅙ pitch and under the graduation figure 12 (representing one-half the span, or the run of the common rafter), find 17, 5, 3, which means 17 feet, 5³⁄₁₂ inches; this is the required length of the hip

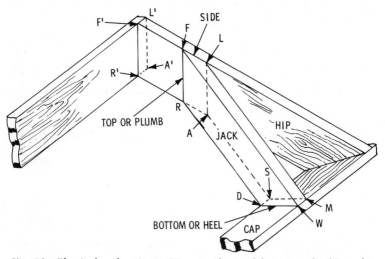

Fig. 30. The jack rafter in position on the roof between the hip rafter and the cap.

or valley rafter. Deduct one-half the thickness of the ridge board, and add for any overhang required beyond the plate.

For the top and bottom cuts of the hip or valley rafter, take 17 on the body of the square and 4 (the rise of the roof per foot) on the tongue. The mark on the 17 side gives the bottom cut; the mark on the 4 side gives the top cut.

For the side cut of the hip or valley rafter against the ridge board, look in the set of figures for the side cut in the table (Fig. 25) under **HIP** along the line for ⅙ pitch, and find the figure 7–7¼. Use 7 on one arm of the square and 7¼ on the other; mark on the 7¼ arm for the side cut.

Reading Length of Rafter per Foot of Run

There are many methods used by carpenters for determining the lengths of rafters, but probably the most dependably accurate method is the "length-per-foot-of-run" method. Since many, perhaps most, of the better rafter-framing squares now have tables on their blades giving the necessary figures, they may almost be considered as standard. The tables may not be arranged in the same manner on all these squares, and on some they may be more complete than on others. The use of these tables is taught in all Carpenter's Union Apprenticeship schools.

Under the heading **LENGTH COMMON RAFTERS PER FOOT RUN** in Fig. 31 will be found numbers, usually from 3 to

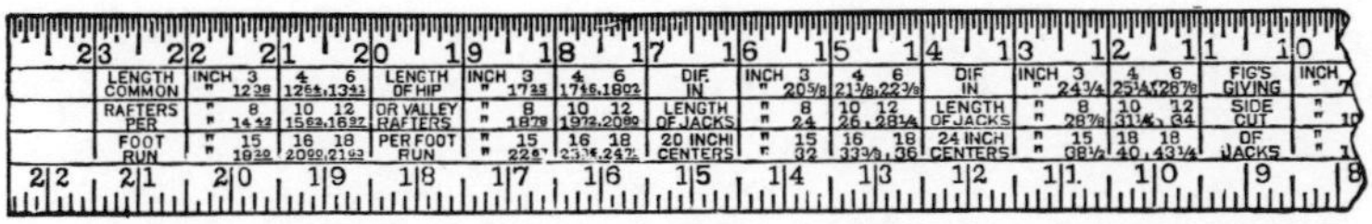

Fig. 31. The length-per-foot-run tables on one type of rafter framing square.

20. With each number is a figure in inches and decimal hundredths. The integers represent the rises of the rafters per foot of run, and the inches and decimals represent the lengths of the rafters per foot of run. As an example of the use of these tables, take a building 28 feet, 2 inches wide, thereby making the run of the rafters 14 feet (allowing for a 2-inch ridge). Let the desired pitch be 4 inches per foot. Under the number 4 on the square will be found the length per foot of run—12.64 inches. The calculation for the length of the rafter is as follows:

$$12.64 \times 14 = 176.96 \ inches$$

259

$$\frac{176.96}{12} = 14.75 \text{ feet, or } 14 \text{ feet, 9 inches}$$

If the run is in feet and inches, it is most convenient to reduce the inches to the decimal parts of a foot, according to the following table:

> 1 inch   equals 0.083 foot
> 2 inches equals 0.167 foot
> 3 inches equals 0.250 foot
> 4 inches equals 0.333 foot
> 5 inches equals 0.417 foot
> 6 inches equals 0.500 foot
> 7 inches equals 0.583 foot
> 8 inches equals 0.667 foot
> 9 inches equals 0.750 foot
> 10 inches equals 0.833 foot
> 11 inches equals 0.917 foot

Example—Find the length of a hip or valley rafter having an 8-inch rise per foot on a 20-foot building with the run of the common rafters measuring 10 feet.

Look for **LENGTH OF HIP OR VALLEY RAFTERS PER FOOT RUN** (Fig. 31), and read under the 8-inch rise the figure 18.76. This is the calculation:

$$18.76 \times 10 = 187.6 \text{ inches}$$

$$\frac{187.6}{12} = 15 \text{ feet, 7.6 inches}$$

One edge of all good steel squares is divided into tenths of inches, so this length may be measured off directly on the rafter pattern with the steel square.

Example—Find the difference in lengths of jack rafters on a roof with an 8-inch rise per foot and with a spacing of 16 inches on centers.

Under **DIFFERENCE IN LENGTH OF JACKS** (16-inch centers) on the square find the figure 19.23 below the figure 8 (rise per foot of run). This is the length of the first jack rafter, and the length of each succeeding jack will be 19.23 inches greater —38.46 inches, 57.69 inch, 76.92 inches, etc.

Example—Find the side cuts of jacks on a square similar to the one shown in Fig. 32.

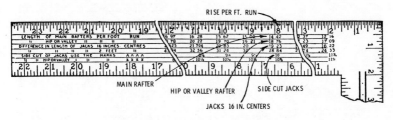

Fig. 32. Typical rafter tables.

The fifth line is marked **SIDE CUT OF JACKS USE THE MARKS** ΛΛΛΛ. If the rise is 8 inches per foot, find the figure 10 under figure 8 in the upper line. The proper side cut will then be 10 × 12, cut on 12. The side cuts for hip or valley rafters are found in the sixth line; for the 8 × 12 roof, it is 10⅞ × 12, cut on 12.

No treatise on rafter framing is complete without an explanation of one of the oldest and most useful, though probably not the most accurate, methods of laying out a rafter with a steel square. Any square may be used if it has legible inch marks representing the desired pitch. It is the same method used for the layout of stairs. Fig. 33 shows the layout for a rafter with a 9-foot run that has a pitch of 7 inches × 12 inches, making the rise of the rafter 5 feet 3 inches. The steel square is applied nine times; carefully mark each application, preferably with a knife. A hip rafter is laid out in exactly the same manner by using 17 instead of 12 in the run and applying the square nine times as was done for the common rafter. For short rafters, this is probably the least time-consuming of any method.

Table of Octagon Rafters

The complete framing square is provided with a table for cutting octagon rafters, as shown in Fig. 34. In this table, the first line of figures from the top gives the length of octagon hip rafters per

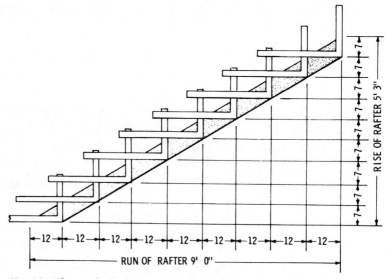

Fig. 33. *The method of stepping off a rafter with a square; the square is applied in consecutive steps, hence the name of the method.*

foot of run. The second line of figures gives the length of jack rafters spaced 1 foot from the octagon hip. The third line of figures gives the reference to the graduated edge that will give the side

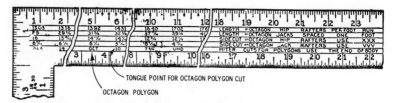

Fig. 34. *Typical octagon rafter tables.*

cut for octagon hip rafters. The fourth line of figures gives the reference to the graduated edge that will give the side cuts for jack rafters. The tables are used in a manner similar to that used for the regular rafter tables just described and therefore, need no further explanation. The last line, or bottom row of figures, gives the bevel of intersecting lines of various regular polygons. At the right end of the body on the bottom line can be read **MITER CUTS FOR POLYGONS—USE END OF BODY.**

262

Example—Find the angle cut for an octagon.

For a figure of 8 sides, look to the right of the word **OCT** in the last line of figures, and find 10. This is the tongue reading; the end of the body is the other point, as shown in Fig. 35.

TABLE OF ANGLE CUTS FOR POLYGONS

This table is usually found on the face of the tongue. It gives the setting points at which the square should be placed to mark cuts for common polygons that have from 5 to 12 sides.

Example—Find the bevel cuts for an octagon.

On the face of the tongue (Fig. 36), look along the line marked **ANGLE CUTS FOR POLYGONS**, and find the reading "8 sides 18–7½." This means that the square must be placed at 18 on one arm and at 7½ on the other to obtain the octagon cut, as shown in Fig. 37.

TABLE OF BRACE MEASURE

This table on the square, shown in Fig. 38, is located along the center of the back of the tongue and gives the length of common braces.

Example—If the run is 36 inches on the post and 36 inches on the beam, what is the length of the brace?

In the brace table along the central portion of the back of the tongue (Figs. 38 and 39), look at L for

$$\begin{matrix} 36 \\ & 50.91 \\ 36 \end{matrix}$$

This reading means that for a run of 36 inches on the post and 36 inches on the beam, the length of the beam is 50.91 inches.

At the end of the table (at F near the body) will be found the reading

$$\begin{matrix} 18 \\ & 30 \\ 24 \end{matrix}$$

This means that where the run is 18 inches one way and 24 inches the other, the length of the brace is 30 inches.

The best way to find the length of the brace for the runs not given on the square is to multiply the length of the run by 1.4142 feet (when the run is given in feet) or by 16.97 inches (when the run is given in inches). This rule applies only when both runs are the same.

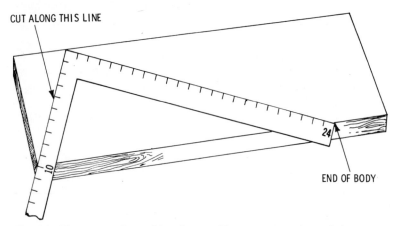

CUT ALONG THIS LINE

END OF BODY

Fig. 35. The square in position for marking an octagon cut; it is set to point 10 on the tongue and to point 24 on the body.

OCTAGON TABLE OR EIGHT-SQUARE SCALE

This table on the square is usually located along the middle of the tongue face and is used for laying off lines to cut an eight-square or octagon-shaped piece of timber from a square timber.

In Fig. 40, let ABCD represent the end section, or butt, of a square piece of 6″ × 6″ timber. Through the center draw the lines AB and CD parallel with the sides and at right angles to each other. With dividers take as many squares (6) from the scale as there are inches in width of the piece of timber, and lay off this square on either side of the point A, such as Aa and Ah; lay off in the same way the same spaces from the point B, as Bd and Be; also lay off Cb, Cc, Df, and Dg. Then draw the lines ab, cd, ef, and gh. Cut off at the edges to lines ab, cd, ef, and gh, thus obtaining the octagon or 8-sided piece.

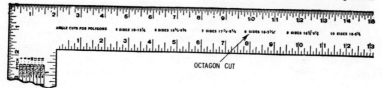

Fig. 36. Table of angle cuts for polygons on the face of the square.

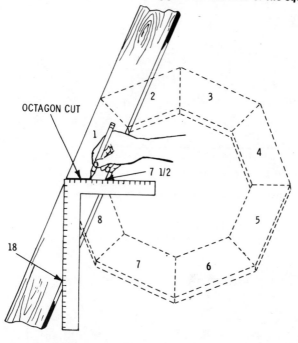

Fig. 37. The application of the square for marking angle cuts of polygons. The square is shown set to points 18 and 7-1/2. When constructing an 8-sided figure, such as an octagon cap, the last figure in the reading is the setting for marking the side. Mark as shown. Cut eight pieces to equal length, with this angle cut at each end of each piece. The pieces will fit together to make an 8-sided figure, as shown by the dotted lines.

ESSEX BOARD MEASURE TABLE

This table is shown in Fig. 41 and normally appears on the back of the tongue on the square. To employ this table, the inch gradua-

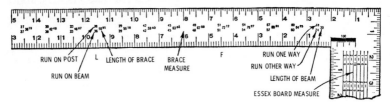

Fig. 38. Table of brace measure on the back of the square.

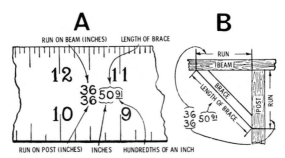

Fig. 39. A portion of the brace-measure table, with an explanation of the various figures, is shown in A. The brace in position, illustrating the measurements of the brace-measure table, is shown in B.

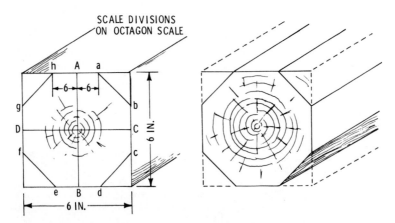

Fig. 40. A square timber and its appearance after it has been cut to an octagon shape, thus illustrating the application of the octagon scale.

tions on the outer edge of the square are used in combination with the values along the five parallel lines. After measuring the length

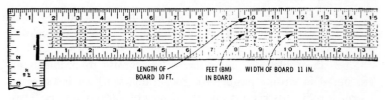

LENGTH OF BOARD 10 FT.   FEET (BM) IN BOARD   WIDTH OF BOARD 11 IN.

Fig. 41. Table of Essex board measure on the back of the square.

and width of the board, look under the 12-inch mark for the width in inches. Then follow the line on which this width is stamped toward either end until the inch mark is reached on the edge of the square where the number corresponds to the length of the board in feet. The number found under that inch mark will be the length of the board in feet and inches. The first number is feet, and the second is inches.

Instead of a dash between the foot and inch numbers, some squares have the inch division continued across the several parallel lines of the scale appearing on one side of the vertical inch division lines and inches on the other.

Example—How many feet Essex board measure in a board 11 inches wide, 10 feet long, and 1 inch thick? 3 inches thick?

Under the 12-inch mark on the outer edge of the square (Fig. 41) find 11, which represents the width of the board in inches. Then follow on that line to the 10-inch mark (representing the length of the board in feet), and find on a line 9–2, which means that the board contains 9 feet 2 inches board measure for a thickness of 1 inch. If the thickness were 3 inches, then the board would contain 9 feet 2 inches $\times$ 3, or 27 feet 6 inches B.M.

SUMMARY

The success of any workshop operation depends on having a sufficient quantity of tools and having a knowledge of their operations. On most construction work, especially in house framing, the steel square is invaluable for accurate measuring and for determining angles.

The square most generally used has an 18-inch tongue and a 24-inch body. The body is generally 2 inches wide, and the tongue

267

is 1½ inches wide, varying in thickness from $\frac{3}{16}$ to $\frac{3}{32}$ of an inch. The various marking on the square are tables and scales or graduations.

Since the tables on the square relate mostly to problems encountered in cutting lumber for roof-frame work, it is necessary to know roof construction and the names of various rafters. These names are rise, run, rise per foot run, hip and valley rafters, jack and cripple rafters, common rafters, ridge and plate. A very good example of these various parts are shown in Fig. 12 of this Chapter.

The rafter tables vary considerably with different makes of squares, not only in the way they are calculated but also in their positions on the square. Some tables are found on the face of the body, and others are on the back of the body. The two general classes of rafter tables found on squares are length of rafter per foot of run, and total length of rafter.

REVIEW QUESTIONS

1. It is called a steel square, but what is the correct name for this tool?
2. What type of tables are found on the body of the square?
3. Name the various types of roof rafters.
4. What is rise per foot run of a roof?
5. What is rafter pitch?

CHAPTER 21

Joints and Joinery

Joinery, or the art of joint-making, is an advanced branch of carpentry and is truly one of the constructive arts. The term "joiner" is seldom used today, but formerly it meant only the most highly skilled woodworkers. It might be said that even reasonably good carpenters and framers were not able to produce the quality joinery required in finely finished buildings, but nearly always the proficient joiner was able to take charge of almost any kind of carpentry.

It may be argued that machine woodworking and factory prefabrication has removed the necessity for skilled hand joinery, but this is not so. As never before, the skilled woodworker has his place. He is the first to be hired when woodworkers are hired, and the last to be laid off when the job is completed, make no mistake about it. Power-operated machines make some of the worst bungled jobs to be found, and many owners are willing, and more than willing, to pay for good, painstaking craftsmanship. It is the contractor who employs good workmen, and not the man who cuts siding and trim with a power handsaw who gets the "cream" of the good work. A thorough knowledge of good joinery and the ability to do it is still the mark of the really skilled carpenter. The Carpenter's Union still bears the euphonious name, the *United Brotherhood of Carpenters and Joiners of America*.

There is no use trying to classify all types of wood joints, because their number and descriptions are infinite, but many of them may be placed under these headings:

1. Straight butt.
2. Dowel.
3. Square, butted, or mitered corners.

4. Dado.
5. Scarf.
6. Mortise and tenon.
7. Dovetail.
8. Wedge.
9. Tongue and groove.

PLAIN EDGE AND BUTTED JOINTS

The plain edge joint is a joint between the edges of boards where the *side* of one piece is placed against the *side* of another, whereas the butt joint is a joint in which the *square end* of one member is placed against the *square end* of another.

Straight Plain Edge Joint

This type of joint is more or less readily made on a power jointer. The plain edge joint has many uses and is commonly used to build up wide boards for panels, shelves, etc., from narrower pieces. For boat planking, the boards are often curved and slightly beveled so that the joint is left open to be calked later. Such curved joints must be fitted one edge to the other. In furniture, cabinet, and other fine finish work, the edges are usually glued.

To make a glued edge joint, square and straighten the edges carefully with both fore and jointer planes. Test the edges often with the try square to assure squareness. Good animal-hide glue is entirely satisfactory for most shop-built assemblies, but it lacks the necessary resistance to dampness and requires a water-jacketed kettle to melt and cook it. Modern polyvinyl glues are used cold; they are easier to use, and they dry quickly in the air. These glues are also nonstaining, and no special equipment is needed for their application. They are now generally used in shops and in furniture factories, though they are not waterproof. If the work is to be exposed to the outside weather, use resorcinol glues; they are entirely waterproof and will air-dry if a catalyst is used, although they are deep red in color and will stain badly. They can be painted over readily. All modern glues will function at room temperatures. If several boards are to be joined edge to edge, as indicated in Fig. 2, at least three clamps will usually be necessary—one on one side and two on the other side—to prevent buckling. Take care to make

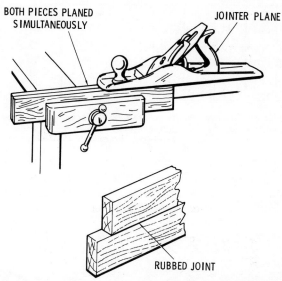

Fig.1. *The method of planing both edges together to obtain a straight-butt side joint. This requires a great deal of skill in planing, and it is necessary that the plane be straight on the edge and carefully sharpened and adjusted. After planing, the edges are glued and rubbed together, as illustrated.*

the edges true even when gluing, or it may unnecessarily require considerable scraping to make the joint flush and smooth.

Dowel Joints

It is usually not necessary to dowel a well-fitting glued edge joint, but it is sometimes done to facilitate assembly; the dowels used are usually quite short. For a butt joint into side wood, they are a satisfactory substitute for mortise and tenon joints and are considerably easier to make. When making heavy screen frames, storm sash, etc., dowel joints are satisfactory if they are glued together with an approved waterproof glue. Fig. 3 shows the assembly of a typical dowel joint.

The holes must be accurately marked and bored; if these precautions are not taken, the holes will not be in perfect alignment and it will be impossible to assemble the joint, or, when assembled, the pieces will not be in their proper alignment. Jigs which are designed to hold the bit in alignment are obtainable from several

271

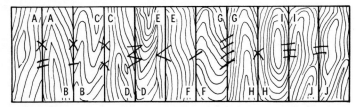

Fig. 2. Narrow boards can be jointed and placed together by using a marking system so that the same edges will come together when assembling them.

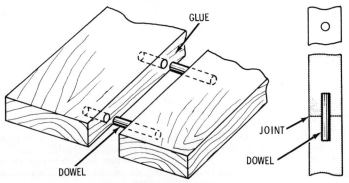

Fig. 3. A typical dowel pin joint.

major tool companies, and these devices are a great help when a great amount of doweling is to be done.

The method of making dowel joints without a jig is shown in Fig. 4. Dowel rods made of several different types of hardwoods are obtainable; some of them have shallow spiral grooves around them to assist carrying the glue into the hole.

Square Corner Joint

The two members of a corner joint are joined at right angles, the end of one butting against the side of the other. When making a corner joint, saw to the squared line with a back saw and finish with a block plane to fit. The work should be frequently tested with a try square, both lengthwise and across the joint. The method of marking this type of joint is shown in Fig. 5. The joint may be fastened together with nails or screws. When fastening the joint, the pieces should be firmly held in position at a 90° angle by a vise or by some other suitable means.

272

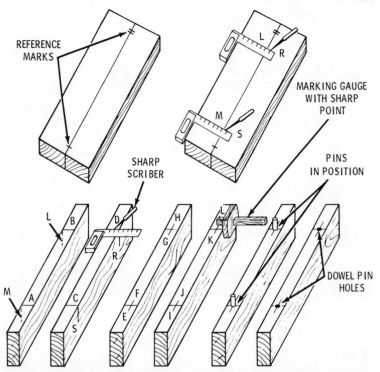

REFERENCE MARKS

SHARP SCRIBER

MARKING GAUGE WITH SHARP POINT

PINS IN POSITION

DOWEL PIN HOLES

Fig. 4. The method of making dowel joints. After making reference marks on the two boards, scribe lines A, B, C, and D. Set the marking gauge to half the thickness of the boards, and scribe lines EF, GH, IJ, and KL. Bore a hole at the intersection of each of these lines; the holes should be just less than half the thickness of the boards. The dowels should fit tightly in these holes.

Mitered Corner Joint

This type of joint is used mostly in making picture frames. To properly make a mitered joint, a picture-frame vise should be used when fastening the pieces together instead of the makeshift method of offset nailing. In fact, a picture-framing shop, to be worthy of the name, should be provided with a picture-frame vise, one type of which is shown in Fig. 6.

When cutting the 45° miter, use a miter box. After sawing, dress and fit the ends with a block plane. There are two ways to nail a mitered joint—the correct way with a picture-frame vise and the

273

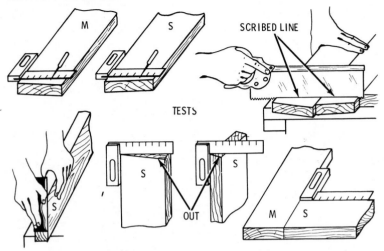

Fig. 5. The method of making a corner joint. After squaring and sawing the edges (M and S), plane the joint surface of one board (S), and test the edges with the square until a perfect right-angle fit is obtained.

wrong way with an ordinary vise. Where considerable work is to be done, a combined miter box and vise is desirable; one of these is shown in Fig. 7.

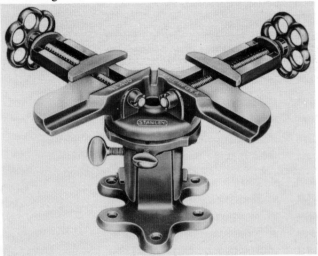

Fig. 6. A typical picture frame vise. With this tool, any frame can be held in the proper position for nailing.

The methods of mitering corners shown in Fig. 8 are used to a great extent when constructing small drawers for merchandise

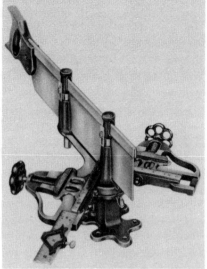

Fig. 7. A typical miter machine. With this device, any type of mitered joint can be cut, glued, and nailed to make tight, close-fitting corners.

cabinets, such as those used in drug stores. The joinery shown in Fig. 8A is not particularly effective. A much stronger joint may be made by sawing the groove for the feather straight across the corner almost through; then glue in strong hardwood feathers, with

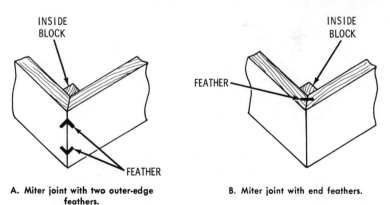

A. Miter joint with two outer-edge feathers.

B. Miter joint with end feathers.

Fig. 8. Miter joints reinforced by feathers. These feathers are kept in place by glue; the joint may also be reinforced by an inside block, as shown.

275

their ends cut off and smoothed flush. The method shown in Fig. 8B is not too efficient and is difficult to make, since the outside corner of the groove is often chipped in construction. These days, the mitered corner joint has been improved by sawing the grooves on a band saw, or on a jigsaw if there are many to make, and then driving a small patented metal feather with sharp turned edges and a slight taper in from each edge. The feathers draw the joint tight, hold well, and no blocking is necessary.

Splined Joint

The form of joint shown in Fig. 9 is called a splined joint, or sometimes a slip-tongue joint. In the shop, it is often used for edge-glued joints, since it holds the members in alignment when clamped.

A groove is made in each of the pieces to be joined, and a spline, made as a separate piece, is inserted in both grooves. The main reason for the use of a splined joint is that when two pieces of softwood are joined, a hardwood spline (which should be cut across the grain) will make the joint less likely to snap than if a tongue were cut in the softwood lengthwise with the grain.

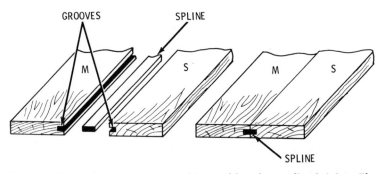

Fig. 9. The component parts and assembly of a splined joint. The spline fits into the grooves in M and S.

Splice Joints

This kind of joint is similar to the familiar double-strap butt joint used on the longitudinal seams of some shell boilers. The two pieces of wood to be joined are placed end to end; they are joined by fish pieces placed on each side and are secured by through-type cross bolts, or nails, as shown in Fig. 10. These fish plates may be

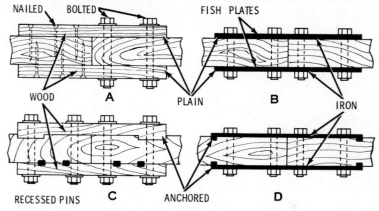

Fig. 10. *Various splice, or fish, joints; A, plain joint with wooden fish plates; B, plain joint with iron fish plates; C, wood plates anchored on the end; D, iron plates anchored on the end.*

made either of wood or of iron, and they may have plain or projecting ends.

The plain type, shown in Fig. 10A, is normally suitable when the form of stress is compression only; however, if the joint is properly made, it will withstand either tension or compression. If the joint is to be subjected to tension, the fish plates (either wood or iron) should be anchored to the main members by keys or projections, as shown in Fig. 10C and D.

LAP JOINTS

In the various joints grouped under this classification, one of the pieces to be joined laps over, or into, the other, hence the name lap joint. Some typical lap joints are shown in Fig. 11.

Housed Butt or Rabbetted Joint

A rabbet is cut across the side of one of the pieces to be joined near the end that is to receive the end of the other piece, as shown in Fig. 12. This form of joint is easily made watertight and is

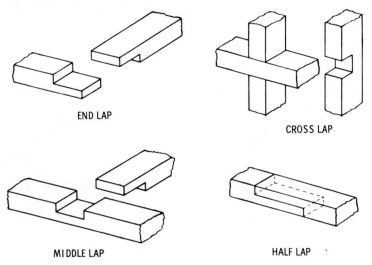

END LAP

CROSS LAP

MIDDLE LAP

HALF LAP

Fig. 11. Typical lap joints. The overlapping feature furnishes a greater holding area in the joint and is therefore stronger than any of the butt or plain joints. A half-lap joint, sometimes called a scarf joint, is made by tapering or notching the sides or ends of two members so that they overlap to form one continuous piece without an increase in thickness. The joints are usually fastened with plates, screws, or nails and are strengthened with glue.

therefore frequently used for tanks and sinks. It is not used where appearance is considered because of the unsightly projecting end.

When building watertight tanks, the rabbets should be accurately cut to size. Both the sides and the bottom are usually carefully selected tank stock, which is practically clear and is obtainable in lengths up to 30 feet and in rather narrow widths, with thicknesses up to 6 inches. The edge joints may be splined but are frequently left square. Wood tanks of this type are still quite popular for the phosphating and acid-pickling processes of steel parts.

Rabbet Joint

A rabbet joint is cut across the edge or end of a piece of stock. The rabbet joint is cut to the width of the stock that will fit into it and to a depth of ½ the thickness of the material. The rabbet joint is a common joint in the construction of cabinet drawers and box construction. (See Fig. 12.)

Dado Joint

A dado joint is a groove cut across the grain and will receive the butt end of a piece of stock. The dado is cut to the width of the stock that will fit in it and to a depth of ½ the thickness of the material. The dado joint is a common joint in construction and is used for the installation of shelves, stairs, and kitchen cabinets.

Scarf Joints

By definition, a scarf joint is made by cutting away the ends of two pieces of timber and by chamfering, halving, notching, or

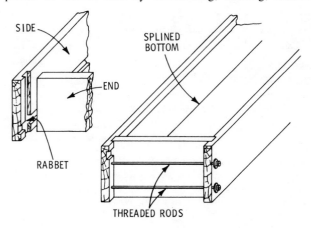

SIDE

SPLINED
BOTTOM

END

RABBET

THREADED RODS

**Fig. 12. The housed, or rabbeted, joint construction
of a wooden pickling tank.**

sloping, making them fit each other without increasing the thickness at the splice. They may be held in place by gluing, bolting, plating, or strapping.

There are various forms of scarf joints, and they may be classified according to the nature of the stresses which they are designed to resist, as:

1. Compression.
2. Tension.
3. Bending.
4. Compression and tension.
5. Tension and bending.

Compression Scarf Joint—This is the simplest form of scarf joint. As usually made, one-half of the wood is cut away from the end of each piece for a distance equal to the lap, as shown in Fig. 13A; this process is called "halving." The length of the lap should be five to six times the thickness of the timber. Mitered ends, as in Fig. 13B, are better than square ends, where nails or screws are depended on to fasten the joint. For extraheavy-duty joints, iron fish plates are sometimes provided, thereby greatly strengthening

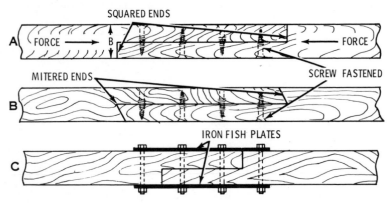

Fig. 13. *Compression scarf joints; A, plain, square ends; B, plain, mitered ends; C, plain, square ends, reinforced with iron fish plates.*

the joint, as shown in Fig. 13C; when these are used, mitered ends are not necessary.

Tension Scarf Joint—There are various methods of "locking" joints to resist tension, such as by means of keys, wedges, or so

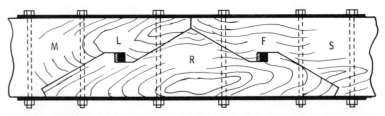

Fig. 14. *Butt and lap plate scarf joint; this joint is designed to avoid reducing the length of the joined timbers when the timbers are not long enough for a lap joint. Piece R is splayed onto timbers M and S, which are cut as shown. The laps of M and S on R are cut with notches and are provided with wedges L and F to handle any tension stress. The joint is bolted and often reinforced with iron fish plates.*

280

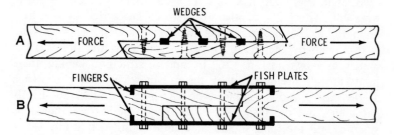

Fig. 15. Tension scarf joints; A, mitered ends fastened with screws; tension stress caused by wedges; B, square ends bolted and reinforced with iron fish plates; tension stress caused by fingers on the fish plates.

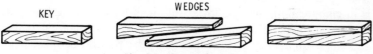

Fig. 16. Key and wedges.

called keys or fish plates with fingers, etc., as shown in Figs. 14 and 15. The difference between keys and wedges, as shown in Fig. 16, should be noted. Keys are only permissible where only one key is

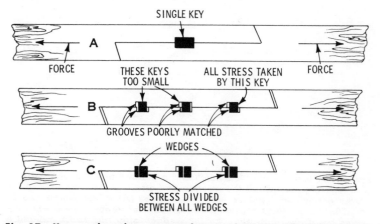

Fig. 17. Keys and wedges are used in scarf joints with mitered ends. If two or more wedges are used in place of one key, they can be made smaller and will provide equal strength to resist the tension stress.

used, as in Fig. 17A; if keys were used where the slots and keys were not perfect fits, all the stress would be carried by one key, as in Fig. 17B, thus rendering the others useless. This problem is overcome by the use of wedges, even when the slots do not match, as shown in Fig. 17C. In modern structural joining, these joints are all strengthened by the use of waterproof glues, such as the casein and resorcinol types. These glues are both slightly gap-filling, but the fitting should be good, the surface should be smooth, and the joint should be clamped until it is firmly set in place.

Bending Scarf Joint—When a beam is acted on by a transverse, or bending stress, the side on which the bending force is applied is subjected to a compression stress, and the opposite side is subjected to a tension stress. Thus, in Fig. 19A, the upper side is in compression, and the lower side is in tension. At L, the end of the joint may be square, but at F, it should be mitered. If this end were square (as at F', Fig. 19B), the portion of the lap M between the bolt and F' would be rendered useless to resist the bending force.

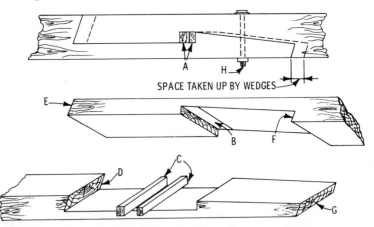

Fig. 18. A scarf joint with a notch and a mitered half lap; the ends are also mitered, illustrating the location and effect of the wedges. The two pieces are joined together with the wedges (A) driven home and cut off. The dotted lines represent the amount of space closed when the pieces are drawn into place by the wedges. Since the cut is mitered at D, E, F, and G, these boards will form a rigid joint, which is often strengthened by a bolt through each section (H).

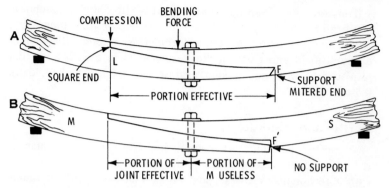

Fig. 19. Bending scarf joints. One end of the joint should be mitered to provide adequate support for the various stresses applied to the joint.

When designing a bending scarf joint, it is important that the thickness at the mitred end be ample, otherwise the strain applied at that point might split the support. Gluing normally helps prevent such stresses from developing.

Compression and Tension Scarf Joint—The essential requirements for this type of joint are flat ends to handle the compression stress and a notched lap with a key, or wedges, to handle the tension stress, as shown in Fig. 20. For severe duty, these

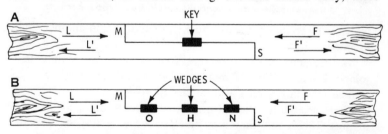

Fig. 20. Compression and tension scarf joints. In A, the key must fit tightly or the compression stress (LF) will be taken by the square lap ends (M and S) and the tension stress (L'F') will be taken by the key. In B, the compression stress (LF) is divided by three wedges (O, H, and N) and the ends (M and S); the tension stress (L'F') is also taken by the three wedges.

joints are sometimes reinforced by iron fish plates with plain or fingered ends. When fish plates with fingers are used, wedges are unnecessary.

Tension and Bending Scarf Joint—This joint is similar to the bending scarf joint in that the lap end of one member is square while that of the other member is mitered. The lap is partly straight and partly inclined, as shown in Fig. 21; a wedge is placed at the middle point against the notches to handle the tension stress.

Mortise and Tenon Joints

A mortise is defined as a space hollowed out in a timber to receive a tenon, and a tenon is defined as a projection, usually with a rectangular cross section, at the end of a piece of timber

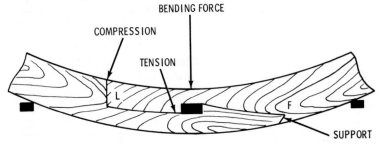

Fig. 21. Tension and bending scarf joint. When making this joint, the square lap end (L) should be on the side which receives the bending force, and the mitered end (F) should be on the other side.

which is to be inserted into a socket, or mortise, in another timber to make a joint.

Mortise and tenon joints are frequently called simply tenon joints. The operation of making mortise and tenon joints is also termed tenoning, which also implies mortising.

There are many different mortise and tenon joints, as illustrated in Figs. 22-28, and they may be classified with respect to:

1. Shape of the mortise
2. Position of the tenon
3. Degree in which tenon projects into mortised member
4. Degree of mortise housing
5. Number of tenons
6. Shape of tenon shoulders
7. Method of fastening the tenon

284

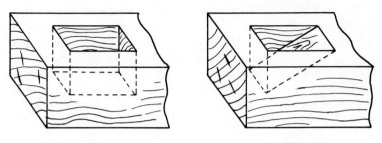

A. Rectangular.   B. Triangular.

Fig. 22. Mortise and tenon joints—shape of the mortise.

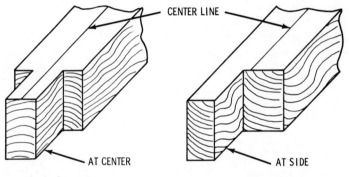

CENTER LINE

AT CENTER   AT SIDE

A. At the center.   B. At the side.

Fig. 23. Mortise and tenon joints—position of the tenon.

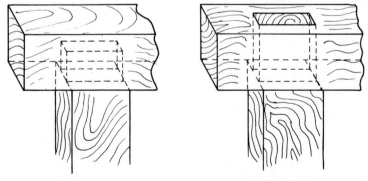

A. Stub tenon.   B. Through tenon.

Fig. 24. Mortise and tenon joints—degree in which the tenon projects into the mortised timber.

285

The mortise and tenon must exactly correspond in size; that is, the tenon must accurately fit into the mortise. The position of the tenon is usually at the center of the timber, but sometimes it is located at the side, depending, except in special cases, on the degree of housing. The tenon may project partly into, or through, the mortised timber. When the tenon and mortise do not extend through the mortised timber, the joint is called a stub tenon. This form of tenon is used for jointing the framework of partitions and is also employed in work where the joint will not be subjected to any tension.

The term "degree of housing" signifies the degree in which the tenon is covered by the mortise, that is, the number of sides of the mortise. The number of tenons depends on the shape of the timbers, whether they are square or rectangular, with considerable width and little thickness, etc. The tenon shoulders are usually at right angles with the tenon as they are when the two timbers are

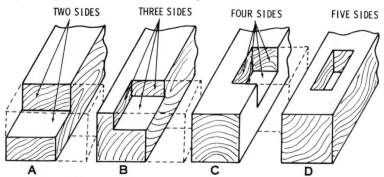

Fig. 25. Mortise and tenon joints—degree of mortise housing; A, two sides; B, three sides; C, four sides; D, five sides.

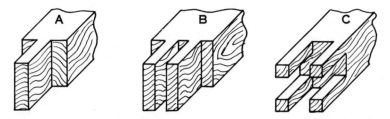

Fig. 26. Mortise and tenon joints—number of tenons; A, single tenon; B, double tenon; C, multitenon.

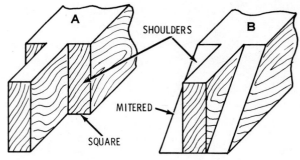

Fig. 27. Mortise and tenon joints—shape of the tenon shoulders; A, square; B, mitered.

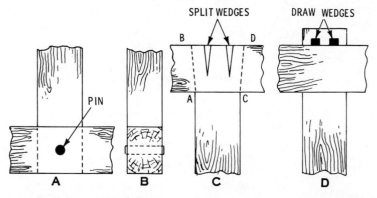

Fig. 28. Mortise and tenon joints—methods of fastening the tenon; A, side view, tenon secured by a pin; B, front view, tenon secured by a pin; C, tenon secured by internal, or split, wedges; sides AB and CD are tapered, thus securely wedging the tenon into the mortise; D, tenon secured by external, or draw, wedges, which are driven into rectangular holes beyond the mortise.

joined at right angles, but they may be mitered to some smaller angle, such as 60° or 45°, as in the case of a brace.

There are several ways of fastening mortise and tenon joints, such as with pins or wedges. When making a mortise and tenon joint, the work is first laid out to given dimensions, as shown in Fig. 29.

Cutting the Mortise—Select a chisel that is as near to the width of the mortise as possible. This chisel, especially for large work, should be a framing or mortise chisel. Bore a hole the same size as the width of the mortise at the middle point. If the

287

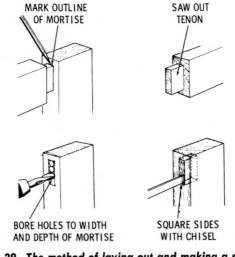

Fig. 29. The method of laying out and making a small mortise and tenon joint.

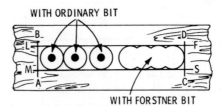

Fig. 30. The method of boring holes when making a large mortise.

mortise is for a through tenon, bore halfway through from each side. In the case of a large mortise, most of the wood may be removed by boring several holes, as shown in Fig. 30. When cutting out a small mortise with a narrow chisel, work from the hole in the center to each end of the mortise, holding the chisel firmly at right angles with the grain of the wood. At the ends of the mortise, the chisel must be held in a vertical position, as shown in Fig. 31B, with the flat side facing the end of the mortise.

Always loosen the chisel by a backward movement of the handle; a movement in the opposite direction would injure the ends of the mortise. Never make a chisel cut parallel with the

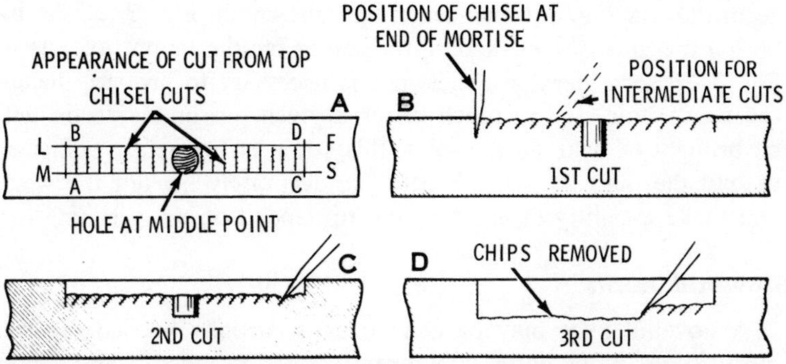

POSITION OF CHISEL AT
END OF MORTISE

APPEARANCE OF CUT FROM TOP

CHISEL CUTS

A | **B**

POSITION FOR
INTERMEDIATE CUTS

1ST CUT

HOLE AT MIDDLE POINT

C | **D** CHIPS REMOVED

2ND CUT

3RD CUT

Fig. 31. The method of cutting a small mortise. After laying out the mortise, bore a hole at the center (A) and work toward each end with a chisel. The chisel cuts should always be made across the grain.

grain, since the wood at the side of the mortise may split. When cutting a through mortise, cut only halfway through on one side, and finish the cut from the other side. After cutting, test the sides of the mortise by using a try square, as shown in Fig. 32; this procedure will check the accuracy with which the work was laid out.

Cutting the Tenon—A back saw is used for cutting out the wood on each side of the tenon, and, if necessary, a finishing cut may be taken with a chisel. After the wood has been cut away, the tenon should be pointed by chiseling all four sides.

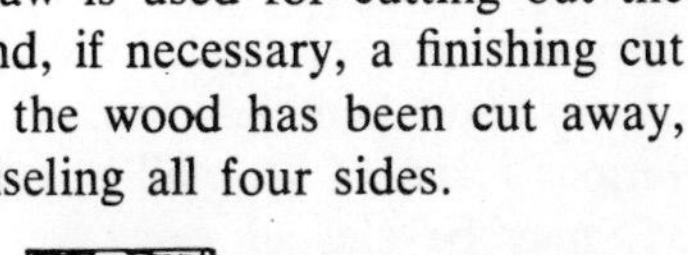

Fig. 32. Test the end with a square after cutting the mortise.

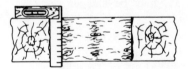

Fig. 33 shows the appearance of the tenon before and after the pointing operation; if this operation were omitted, a tight-fitting tenon would be difficult to start into the mortise and could splinter the sides of the mortise when driven through. Do not cut off the point until the tenon is finally in place and the pin is driven home.

Draw Boring—The term "draw boring" signifies the method of locating holes in the mortise and tenon that are eccentric with each other so that when the pin is driven in, it will "draw" the tenon into the mortise, thereby forcing the tenon shoulders tightly against the mortised timber. The holes may be located either by accurately laying out the center, as shown in Fig. 34C, or by boring the mortise and finding the center for the tenon hole, as in Fig. 34D. Considerable experience is necessary to properly locate the tenon hole. If too much offset is given, an undue strain will be brought to bear on the joint; this strain is frequently sufficient to split the joint. It is much better to accurately lay out the work and make a tightly fitting pin than to depend on draw boring.

Dovetail Joints

A dovetail joint may be defined as a partially housed tapered mortise and tenon joint, the tapered form of mortise and tenon

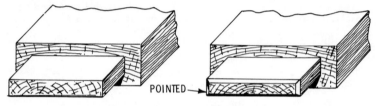

Fig. 33. Appearance of the tenon before and after pointing.

forming a lock which securely holds the parts together. The word "dovetail" is used figuratively; the tenon expands in width toward the tip and resembles the fan-like form of the tail of a dove. The various forms of dovetail points, some of which are shown in Fig. 35, may be classed as:

1. Common
2. Compound
3. Lap, of half-blind
4. Mortise, or blind.

Common Dovetail Joint—This is a plain, or single "pin," joint. In dovetail joints, the tapered tenon is called the *pin,* and the mortised part that receives this joint is called the *socket*. Where strength rather than appearance is important, the common dovetail joint is used. The straight form of this joint is shown in Fig. 36, and the corner form is shown in Fig. 37; the proportions of the joint are shown in Fig. 37A and B.

Compound Dovetail Joint—This is the same as the common form but has more than one pin, thereby adapting the joint for use

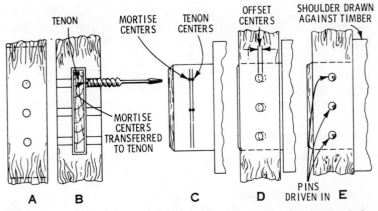

TENON · MORTISE CENTERS · TENON CENTERS · OFFSET CENTERS · SHOULDER DRAWN AGAINST TIMBER · MORTISE CENTERS TRANSFERRED TO TENON · PINS DRIVEN IN

A   B   C   D   E

Fig. 34. The method of transferring pin centers from the mortise holes to the tenon by draw boring. When laying out the tenon-hole centers, make the offset toward the tenon shoulder.

with wide boards. When making this joint, both edges are made true and square; a gauge line is run around one board at a distance from the end equal to the thickness of the other board, and the other board is treated similarly. Two methods are commonly followed. Some mark and cut the pins first; others mark and cut the sockets first.

In the first method, the pins are carefully spaced, and the angles of the tapered sides are marked with the bevel. Saw down to the gauge line, and work the spaces in between with a chisel and a mallet. Then, put B on top of A (in Fig. 38), and scribe the mortise. Square over, cut down to the gauge line, clean out, and fit together.

The second method is to first mark the socket on A (sometimes on common work, the marking is dispensed with, and the worker uses his eyes as a guide); then, run the saw down to the gauge line, put A on B, and mark the pins with the front tooth of the saw. Cut the pins, keeping outside of the saw mark sufficiently to allow the pins to fit tightly; both pieces may then be cleaned out and tried together.

When cleaning out the mortises and the spaces between the pins, the woodworker must cut halfway through, then turn the board over and finish from the other side, taking care to hold the

291

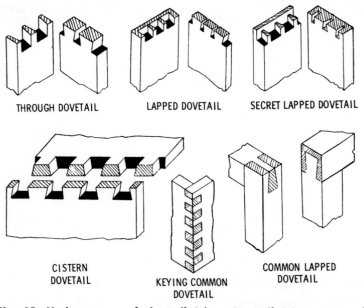

THROUGH DOVETAIL   LAPPED DOVETAIL   SECRET LAPPED DOVETAIL

CISTERN DOVETAIL   KEYING COMMON DOVETAIL   COMMON LAPPED DOVETAIL

Fig. 35. Various types of dovetail joints. Dovetail joints are used principally in cabinetmaking, drawer fronts, and fine furniture work. They are a partly housed and tapered form of tenon joint in which the taper forms a lock to hold the parts securely together.

chisel upright so as not to undercut, as shown in Fig. 38, which is sometimes done to insure the joint fitting on the outside.

Lap or Half-Blind Dovetail Joint—This joint is used in the construction of drawers on the best grades of work. The joint is visible on one side but not on the other, as shown in Fig. 39, hence the name "half blind." Since this form of dovetail joint is used so extensively in the manufacture of furniture, machines have been devised for making the joint, thus saving time and labor.

Blind Dovetail Joint—This is a double lap joint; that is, the joint is covered on both sides, as shown in Fig. 39, and is sometimes called a secret dovetail joint. The laps may be either square, as in Fig. 40, or mitered, as in Fig. 41. Because of the skill and time required to make these joints, they are used only on the finest work. The mitered form is the more difficult of the two to assemble.

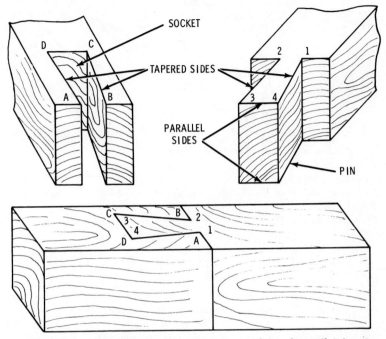

Fig. 36. The straight form of the common, or plain, dovetail joint. By noting the positions of the letters and numbers, it may be seen how the socket and pin are assembled.

Spacing—The maximum strength would be gained by having the pins and sockets equal; however, this is rarely done in practice, since the mortise is made so that the saw will just clear at the narrow side with the space from eight to ten times the width of the widest side. Small pins are used for the sake of appearance, but fairly large ones are preferable. The outside pin should be larger than the others and should not be too tight or there will be the danger of splitting, as shown in Fig. 42 at point A. The angle of taper should be slight (70° to 80°) and not acute as shown, otherwise there is the danger of pieces L and F in Fig. 42 being split off in assembling.

Position of Pins—When boxes are made, the pins are generally cut on the ends with the sockets on the sides. Drawers have the pins on the front and back. The general rule is to locate the

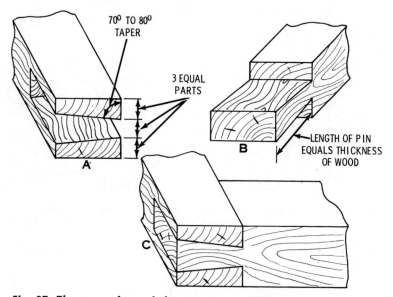

70⁰ TO 80⁰ TAPER

3 EQUAL PARTS

LENGTH OF PIN EQUALS THICKNESS OF WOOD

A

B

C

Fig. 37. *The corner form of the common, or plain, dovetail joint, with the proper proportions for the socket and pin.*

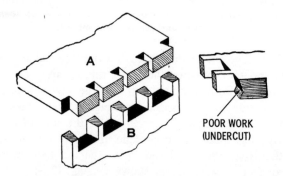

A

B

POOR WORK (UNDERCUT)

Fig. 38. *A multiple dovetail joint, with a poorly cleaned joint shown in detail.*

tapered sides so that they are in opposition to the greatest stress that may be applied on the piece of work to which the joint is connected.

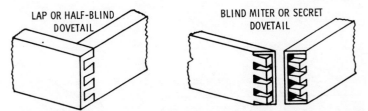

Fig. 39. Half-blind and blind dovetail joints, respectively. These joints are used in the best grades of drawer and cabinetwork, since the joint is visible on only one side. They should be exceptionally well fitted because of the frequent pull on the front piece.

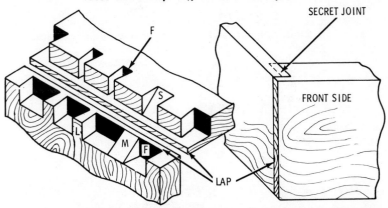

Fig. 40. The blind, square-lap dovetail joint. Two forms of pins and sockets are used—mitered (MS) and square (LF).

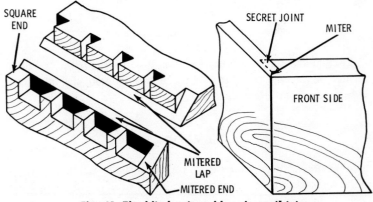

Fig. 41. The blind, mitered-lap dovetail joint.

295

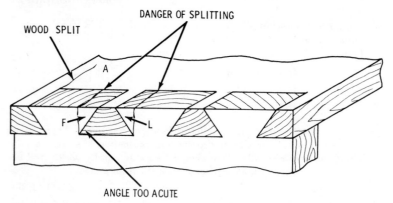

Fig. 42. *A badly proportioned common dovetail joint can result in splitting.*

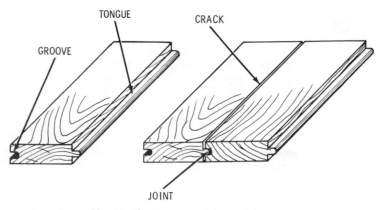

Fig. 43. The tongue-and-grove joint.

Tongue-and-Groove Joint—In this type of joint, the tongue is formed on the edge of one of the pieces to be joined, and the groove is formed in the other, as shown in Fig. 43.

SUMMARY

In woodworking, the term "joint" means the union of two or more smooth or even surfaces permitting a close fitting or junction, as in a joint between two boards. The aim is to obtain a strong joint without weakening any part unduly by the removal of too much wood.

There are many wood joints, all of which may be divided into the following classifications; plain or butt joints, lap joints, mortise-and-tenon joints, and dovetail joints, just to name a few. A plain or butt joint is where the end or one side of a piece is placed or butted against one end or side of the other. A lap joint is where two pieces to be joined lap over or into one another.

Plain or butt joints are generally classified as straight or plain-edge, dowel pin, splined or feather, beveled spline or miter, and beveled plain edge. Straight or plain-edge joints are the simplest form of joint and has many uses where several pieces are required to form a flat surface.

Dowel joints could be considered as a substitute for mortise-and-tenon joints. If well made and not exposed to weather and extreme temperature changes, it is a strong and excellent joint. A dowel joint is simply a butt joint reinforced by dowels which fit tightly into holes bored in each member to align them with each other.

There are many variations of the mortise-and-tenon joint used in cabinetwork. Among the more common joints are stub tenon, through tenon, haunched tenon, open tenon, and double tenon. When cutting a mortise, select a chisel as near the width of the mortise as possible. Check all cuts for accuracy before applying glue.

REVIEW QUESTIONS

1. What is a plain or butt joint?
2. What is a mortise-and-tenon joint?
3. What is a dovetail joint?
4. What are dowel joints? How do they compare with mortise-and-tenon joints?
5. Name the five mortise-and-tenon joints.

CHAPTER 22

Cabinetmaking Joints

The purpose of this chapter is to describe and illustrate some of the many joints and their applications as used in cabinetwork in the hope that this information may be found useful to carpenters and others when the occasion for such information arises.

THE TOOLS

A full set of good carpenters' tools is necessary in the cabinet-making shop, including a set of firmer chisel, a set of iron bench planes, a set of auger bits with slow-feed screws that range in size from $\frac{1}{4}$ to 1 inch, and a set of numbered bits for the electric drill. In addition, the following tools will be found useful almost continually.

1. Router plane
2. Plow plane
3. A substantial cast-iron miter box
4. Several bar clamps with varying lengths of bars
5. Hand clamps (those with wood jaws are most useful in the shop, but malleable C-clamps are often used)
6. A high-speed $\frac{1}{4}$-inch electric drill
7. A chute board of wood or iron.

Most of these tools have been described previously in this book.

THE BENCH

A workbench with an end vise, in addition to the regular vise, for holding material between stops on top of the bench will be found quite convenient.

JOINTS

The great variety of joints used in cabinetwork are usually classified according to their general characteristics, as glued, halved and bridle, mortise and tenon, dovetail, mitered, framing, and hinging and shutting. Under each of these classifications is grouped a variety of joints which will be considered separately and will also be briefly explained.

Glued Joints

In cabinetwork, practically all joints are, or should be, glued. There are several types of glues suitable for use in the cabinet shop.

For large, continuous, and high-speed assemblies, the modern synthetics are used, but most of them are thermo-setting; that is, they must be hot-pressed, and they require elaborate and expensive equipment. For medium to large operations, especially structural joinery, the natural caseins seem to stand alone. They are reasonably economical, work at room temperatures, and are highly water-resistant and gap-filling; however, they are highly alkaline and stain some woods badly and, when dry, are so hard that they will quickly take the edges off of tools and machine blades.

The resorcinols are excellent adhesives, and joints made with these glues will even withstand weathering, but they are a deep red color which makes them unsuitable for some uses. They have great strength and, with a catalyst, can be set up at room temperatures.

The glue which seems to be the most popular in woodworking shops is the polyvinyls. They are nonstaining, light colored, easy to use, have long pot life, and can be set at room temperatures. They do not require heating; however, they are not waterproof to any great degree. Glue technology is changing rapidly, and there are so many formulations available that it is not possible to make any recommendations for specific jobs. All glued joints must be clamped, or pressure must be applied by means of nails or screws. This is vital. Good glued joints must be in intimate contact while they are setting up. If this contact is not made, the joint will not have the strength required of it.

Beveled Joints—In this type of joint, the sides of the pieces fit together to form angles, or corners, as shown in Fig. 1. An infinite amount of planing and dressing can be saved by first ripping the edges roughly, by hand if necessary, on a tilting-arbor table saw if one is available. If ripped on a power saw, the angle can be adjusted precisely, and only a small amount of hand dressing will be necessary. Try the bevel continually with a T-bevel while dressing to assure that the joints fit; they must

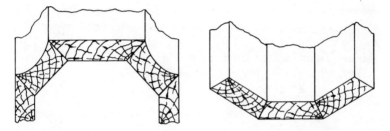

Fig. 1. Typical beveled joints.

fit properly if the joint is to be glued. The joints must be clamped, and without special clamps this is troublesome. The woodworker's ingenuity will usually suggest a method. Short pieces of chain with bolts through the end links are useful, if the chains can be passed around the work. Beveled joints do not usually require exceedingly high pressure.

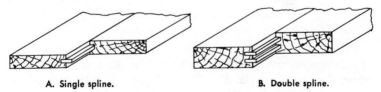

A. Single spline.   B. Double spline.

Fig. 2. Single- and double-splined joints.

Plowed-and-Splined Joints—This method of jointing is commonly used in cabinetwork and is similar to the spline joint in the preceding chapter, except that in cabinetwork, the splines are cut *across* the grain. When the thickness of the material will permit, two splines are used instead of one, as in Fig. 2B, because of the additional gluing surface afforded and the increased strength to the joint. Splines are cut lengthwise with the grain.

To make this joint successfully, the pieces should be properly faced, and the edges should be squared and straightened with a jointer so that they fit perfectly. Put reference marks on the face side so that the same edges will come together when assembled. Set the plow plane with the iron projecting approximately $\frac{1}{32}$ inch below the bottom plate. Set the depth gauge to one-half the width of the spline, and adjust the fence so that the cutter will be the required distance from the edge between the two sides. Fasten the piece securely in the bench vise so that the groove can be plowed from the face side. Begin plowing at the front (Fig. 3A) and work backward; finish by going right through from back to front, as in Fig. 3B. Hold the plow plane steady, otherwise an irregular groove will result.

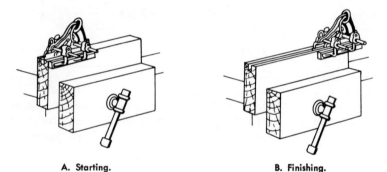

A. Starting. B. Finishing.

Fig. 3. The method of plowing a single spline groove.

For the cross-grain spline, cut off the end of a thin board of hardwood; mark it, and carefully saw off a strip across its width that is the required width of the spline, approximately $\frac{3}{4}$ inches wide. Plane the spline to the desired thickness in a tonguing board. Then, assemble the parts and glue them up, as shown in Fig. 4.

Hidden Slot Screwed Joints—This joint is not often used as a glued joint but is found to be an effective way of fastening brackets and shelves to finished work where the fastening must

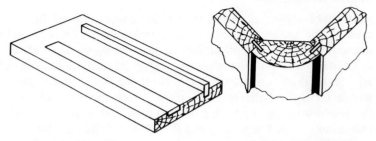

A. A tonguing board.   B. Plowed-and-splined joints.

Fig. 4. The tonguing board is a simple and handy device when used to overcome the difficulty of holding a narrow piece of thin material steady while planing. To make the board, use a piece of 7/8-inch faced material, 8 to 10 inches wide and longer than the tongues to be planed. Cut the grooves, as indicated, with a tenon saw; clean out the grooves with a chisel and a router. The wider groove should be slightly deeper than the thickness of the finished tongue to allow for planing both sides of boards placed in it.

be concealed. The joint consists of a screw which is driven part way into one piece and a hole and slot cut into the opposite piece. The joint is effected by fitting them together with the head of the screw in the hole and then forcing the screw back into the slot. Fig. 5A shows the slot and screw in relation to each other, and Fig. 5B gives a cross-sectional view of the completed

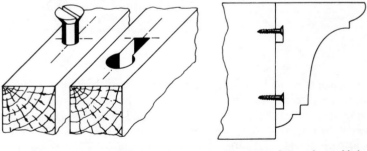

A. The joint before assembly.   B. Cross-sectional view of assembled joint.

Fig. 5. The hidden slot screwed joint.

joint. This joint is also used in interior work for fastening pilasters and fireplaces to walls, for panelling, and for almost every kind of work requiring secure and concealed fastening.

303

To make a joint as shown in Fig. 5B, gauge a center line on each of the pieces. Determine the position of the screws and insert them; they should project approximately ⅜ inch above the surface. Hold the two pieces evenly together, and, with a try square, draw a line from the back of the screw shank across the center line of the opposite piece. From this line, measure ⅞ inch forward on the center line; with this point as the center, bore a hole to fit the screw head that is just slightly deeper than the amount that the screw projects above the surface. Cut a slot from the hole back to the line from the screw shank; this slot should be as wide as the diameter of the shank and as deep as the hole. As a general rule, the total length of the slot and hole should be slightly more than twice the diameter of the head of the screw.

The process of fastening pilasters to fireplaces by this method is as follows: First, mark the position of the piece and the place on the wall for the screws. In brick and cement walls, holes are drilled and wooden plugs are driven in flush with the surface to hold the screws. The plugs are shaped as illustrated in Fig. 6. Plugs cut as shown seldom work loose. Turn the screws

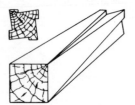

Fig. 6. A plug is used to hold the screw when hidden slot screwed joints are utilized on mortar and brick walls.

into the plugs, and allow them to project approximately ⅜ inch from the surface; the screw heads should be smeared with moist lampblack. Put the piece in position and press it against the screw heads; this pressure will leave black impressions. Bore holes to fit the screw heads approximately ⅝ inch below the impressions, and cut the slot to receive the shank of the screw. Replace the piece, with the heads in the holes, and force it down.

Dowel Joints—There are many variations of dowel jointing commonly used in cabinetwork. The basic principle and method of making a dowel joint is explained and illustrated in the preceding chapter.

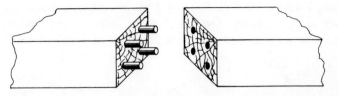

Fig. 7. The use of dowels in a butt joint adds strength to the joint. This type of construction is frequently found in cabinet work to lengthen large mouldings and where the cross grain prevents tenoning.

To accurately fix the position of dowels in a butt joint, such as the one shown in Fig. 7, make all measurements and gauge lines from the edge of the faced sides. For example, with material that is 4 inches square, mark diagonal lines from the corners with a scratch awl, intersecting at the center. Then, from the edge of the faced sides, mark off 1 inch and 3 inches as shown in Fig. 8A. From the same sides, gauge the lines as in Fig. 8B. The intersection of the lines is the center for the holes, as shown in Fig. 8C.

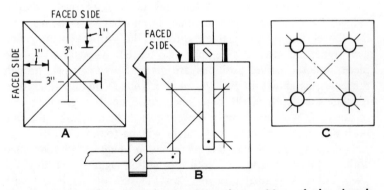

Fig. 8. The method used for marking the position of the dowels in a butt joint.

Where the ordinary means of aligning dowel holes cannot be used, a dowel template or pattern is used. The template is usually made of a strip of zinc or plywood, with a small block of wood fastened to one end to act as a shoulder; the position of the dowel points is then pierced through the zinc or plywood pattern with a fine awl. Various types of templates are made and used as the occasion requires. Fig. 9A shows a template that is used for making dowel rails in furniture; it is made to fit the

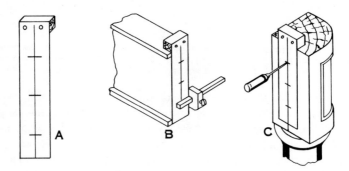

Fig. 9. The use of a template, or pattern, for marking dowel-pin locations; A, the template; B, the template is made to fit the section of rail; C, marking dowel positions on leg.

section of the rail in Fig. 9B. While held in position, a line is gauged down through the middle; the position of the dowels is indicated on the line. The template is laid flat on a board, and the dowel points are pierced through the surface with a fine awl. When in use, the template is placed in position on the piece, and the dowel positions are marked with an awl through the holes, as shown in Fig. 9C. A "bit gauge" should be used to regulate the depth of the bore when doweling. If a great amount of doweling is to be done, a doweling jig, which insures accurate boring of holes from $\frac{1}{4}$ to $\frac{3}{4}$ inch, will be found useful.

Dowels are glued into one piece, cut to length, and sharpened with a dowel sharpener. As a precaution against splitting the joint, cut a V-shaped groove down the side of the dowel with a chisel; this groove permits the glue and air to escape.

Coopered Joints—These are so named because of the resemblance to the joints used in barrels made by coopers, and are used for practically all forms of curved work; they are usually splined before gluing, although dowels are used occasionally. Fig. 10 shows the coopered joint in semicircular form, with the segments beveled at an angle of 15 degrees. They are clamped after gluing and planed to shape.

Halved and Bridle Joints

These are lap joints with each of the pieces halved and shouldered on opposite sides, so that they fit into each other. They are the simplest joint used in cabinetwork. Fig. 11A shows the common halved angle, which is the one most frequently used. Fig. 11C illustrates the oblique halved joint, which is used for oblique connections. Fig. 11D represents the mitered halved joint, which is useful when the face or frame piece is moulded. Fig. 11E, F, and G shows the joints that are used for cross connections having an outside strain. Fig. 11H illustrates the blind dovetail halved joint, which is used in places where the frame edge is exposed. Bridle or open-tenon joints are used to connect parts of flat and moulded frames. The joint in Fig. 12B is used where a strong framed groundwork, which is to be faced up, is required. The joint in Fig. 12D is used as an inside frame connection.

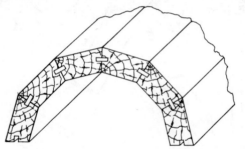

Fig. 10. Coopered joints are employed to form various curvatures in cabinet work.

Mortise and Tenon Joints

Many variations of the mortise and tenon joints are used in cabinetwork. They differ in size and shape according to the requirements of the location and the purpose for which the joint is used. The most frequently used mortise and tenon joint is the stub tenon, so called because it is short and penetrates only part way through the wood; Fig. 13A illustrates the type most generally used in doors and furniture framing. Fig. 13B shows the stub tenon with a mitered end; this type of construction is often

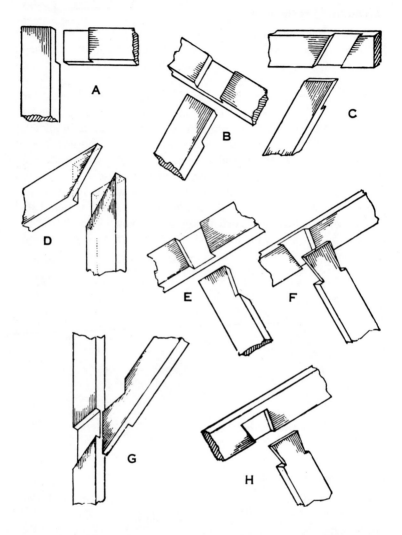

Fig. 11. Various halved joints; A, halved angle joint; B, halved tee joint; C, oblique halved joint; D, mitered halved joint; E, dovetail halved joint; F, dovetail halved joint; G, oblique dovetail halved joint; H, blind dovetail halved joint.

necessary when fitting rails into a corner post. Fig. 13C shows the rabbeted or "haunched" tenon, which is considered a stronger

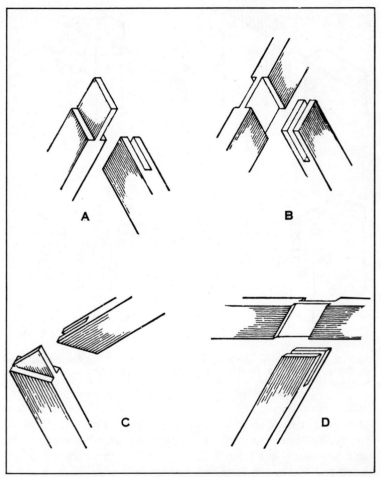

Fig. 12. Halved joints that are generally used to make flat and molded frames; A, angle bridle joint; B, tee bridle joint; C, mitered bridle joint; D, oblique bridle joint.

joint because of the small additional tenon formed by the rabbet; it is often mitered, as shown in Fig. 13D, to conceal the joint when used on outside frames. The joint in Fig. 14A is the same as that shown in Fig. 13C, but it is shouldered on one side only; it is sometimes called a "barefaced" tenon and is used when the connecting rail is thinner than the stile into which it is joined. Fig. 14B shows the long and short shoulder tenon; this joint is used

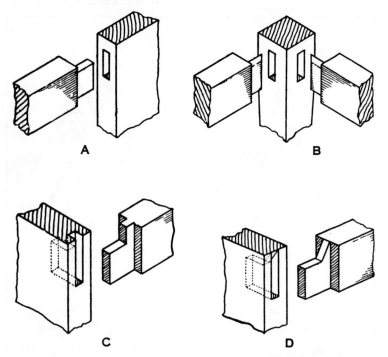

Fig. 13. Variations of the mortise and tenon joints are frequently used in doors and in the framing of furniture; A, stub mortise and tenon; B, mitered stub tenon; C, haunched, or rabbeted, mortise and tenon; D, haunched and mitered mortise and tenon.

when connecting a rail into a rabbeted frame, since it has one shoulder cut back so as to fit into the rabbet. Fig. 14C illustrates the double tenon joint, which increases the lateral strength of the stile into which it is jointed. It is simply a stub tenon that is rabbeted and notched to form two tenons, and, when glued, it makes an exceptionally strong joint. Fig. 14D represents a type of through mortise and tenon that is sometimes used for mortising partitions into the top or bottom of wardrobes, cabinets, etc.; the partitions are wedged across the tenon and glued.

Laying Out the Mortise and Tenon—The general practice when laying out a mortise and tenon is to square the mortise lines across the edge of the stile in pencil and then scribe two lines for the sides of the mortise with a mortise or slide gauge between the pencil lines. If the tenon is to be less than the full

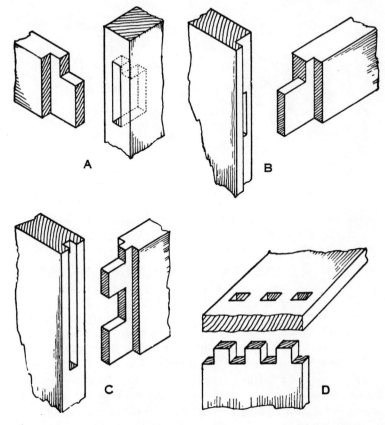

Fig. 14. *Other mortise and tenon joints used in the construction of furniture; A, barefaced mortise and tenon; B, long and short shoulder mortise and tenon; C, double mortise and tenon; D, pinned mortise and tenon.*

width of the rail, square the rail lines across the edge, in addition to the mortise lines, as shown in Fig. 15A. This procedure insures greater accuracy when designating the position of the mortise. When two or more stiles are to be mortised, they are clamped together, and the lines are squared across all the edges simultaneously.

For a through mortise, continue the pencil lines across the face side and onto the back edge; gauge the mortise lines from the faced side. With the gauge set for the mortise, scribe the lines

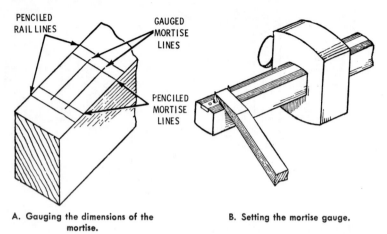

A. Gauging the dimensions of the mortise.

B. Setting the mortise gauge.

Fig. 15. Laying out the mortise.

for the tenon on both edges and the end of the rail, and with the aid of a try square, mark the shoulder lines with a knife or chisel on all four sides.

The proportions of stub and through mortises and tenons are usually considered as about ⅓ the thickness of the wood, and they should be cut with a mortise chisel of the required size. If the chisel is not exactly ⅓ the thickness of the material, it is better to make the mortise more than ⅓ rather than less. Set the mortise gauge so that the chisel fits exactly between the points, as shown in Fig. 15B. Make a chisel mark in the center of the edge to be mortised, and adjust the head of the gauge so that the points coincide with this mark.

Mortise cutting in cabinetwork is usually done entirely with a mortise chisel, beginning at the center and working toward the near end with the flat side of the chisel toward the end. Remove the core as you proceed; then reverse the chisel, and cut to the far end, being careful to keep the chisel in a perpendicular position when cutting the ends. Through mortises are cut halfway through from one side, and the material is then removed and cut through from the opposite side.

A depth gauge for stub mortises is made by gluing a piece of paper or tape on the side of the chisel, as shown in Fig. 16. If the method of boring a hole in the center from which to begin the cutting of the mortise is used for stub mortises, it is advisable to

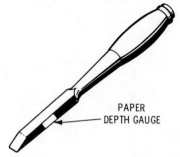

Fig. 16. The depth of the mortise joint may be controlled by fastening a piece of paper or tape on the side of the chisel.

PAPER
DEPTH GAUGE

use a bit gauge to regulate the depth of the bore. A small firmer chisel is used to clean out stub mortises.

Cutting the Tenon—Fasten the piece firmly in tne bench vise. Start the cut on the end grain, and saw diagonally toward the shoulder line, as in Fig. 17A. Finish by removing the material in the vise and cutting downward flush with the edge, as in Fig. 17B. The diagonal saw cut acts as a guide for the finishing cut and provides greater accuracy. Small tenons are usually cut with a dovetail saw.

Cutting the Shoulder—After making the tenon cuts, and to overcome any difficulty in cutting the shoulders, place the piece on the shoulder board or bench hook, and carefully chisel a V-

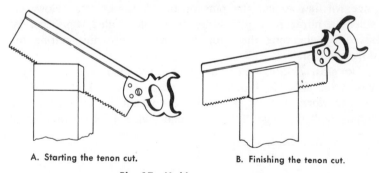

A. Starting the tenon cut.

B. Finishing the tenon cut.

Fig. 17. Making a tenon cut.

shaped cut against the shoulder line, as shown in Fig. 18A. Hold the work firmly against the stop on the board; place the saw in the chiseled channel, and begin cutting by drawing the saw backward and then pushing it forward with a light stroke. Hold the thumb and forefinger against the saw, as in Fig. 18B, and keep

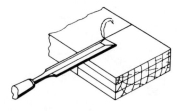

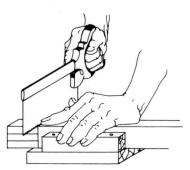

A. Starting the shoulder cut.

B. Keep the saw upright while cutting the shoulder.

Fig. 18. Making the shoulder cut.

the saw in an upright position. A straightedge can be placed against the shoulder line to act as a guide when cutting wide shoulders. In the case of extremely wide tenons and shoulders, a rabbet plane and a shoulder plane are used; the straightedge is used as a guide for the rabbet plane.

Dovetail Joints

The method of making dovetail joints is described in the previous chapter. In common dovetailing, it is a matter of convenience whether to cut the pins or the dovetails first. However, where a number of pieces are to be dovetailed, time can be saved by clamping them together in the vise and cutting the dovetails first.

Dovetail Angles—For particular work where the joint is exposed, the dovetails should be cut at an angle of 1 in 8, and for heavier work, 1 in 6. To find the dovetail angle, draw a line square with the edge of a board, and divide it into 6 or 8 equal

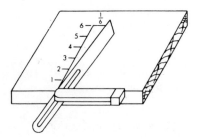

Fig. 19. Finding the angle of the dovetail.

parts as desired; from the end of the line and square with it, mark off a space equal to one of the divisions, and set the bevel as shown in Fig. 19.

A dovetail template, as shown in Fig. 20A, will be found quite handy if there is a great deal of dovetailing to be done. To make the template, take a rectangular piece of ¾-inch material of any desired size, and square the edges; with the mortise gauge set for a ¼-inch mortise at ¼ inch from the edge, scribe both edges and one end. With the bevel set as shown (Fig. 19), mark the shoulder lines across both sides of the lower portion, and cut it with a tenon saw. Make one cut for each of the two angles. The template may also be made by gluing a straightedge, at the required angle, across both sides at one end of a straight piece of thin material. The use of a dovetail template saves time and insures uniformity. Place the shoulder of the template against the edge, as shown in Fig. 20B, and mark one side of the dovetail along its edge. Reverse the template, place the other shoulder at the same edge, and mark the other side of the dovetail.

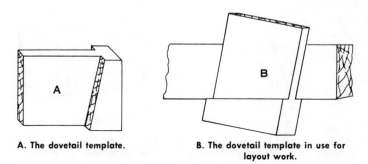

A. The dovetail template.

B. The dovetail template in use for layout work.

Fig. 20. A template is invaluable for dovetail work.

Beveled Dovetailing—The joint shown in Fig. 21 is frequently required in cabinetwork, and a template is a great help for marking it. To use the template for marking beveled dovetails, cut a wedge-shaped piece of material, as shown in Fig. 22A, that is beveled at the same angle as the bevel of the material to be dovetailed. Insert this wedge between the edge of the material and the template, with the square edge of the wedge against the shoulder of the template, as in Fig. 22B. Mark the dovetail as described, but do not reverse the wedge-shaped piece.

The common, or through dovetail, shown in Fig. 23A, is primarily used for dovetailing brackets and frames which are subject to a heavy downward strain. Fig. 23B illustrates the common lapped, or half-blind, dovetail as it is applied to a curved door frame; it is used in all locations of this type where

Fig. 21. The beveled dovetail joint.

mortise and tenon joints would not be effective. The common lapped dovetail joint may also be used for purposes similar to those described for the common dovetail joint.

A

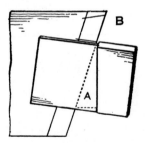

B

A

A. A wedge is used with the dovetail template to mark the desired bevel.

B. Laying out beveled dovetails with the template and wedge.

Fig. 22. The method of laying out work with the aid of a dovetail template.

Fig. 24A illustrates the common housed "bareface" dovetail; it is shouldered on one side only. The joint in Fig. 24B is shouldered and dovetailed on both sides and is another of the same type with the dovetailing parallel along its entire length. These are the simplest forms of housed dovetailing. Their application to the framing of furniture is shown in Fig. 29, under the section on framing joints discussed later in this chapter.

Fig. 24C illustrates a shouldered housing dovetail joint with the dovetail tapering along its length; as with the two preceding joints, this joint can be shouldered on one side or both sides. The tapered dovetail makes this joint particularly adaptable for con-

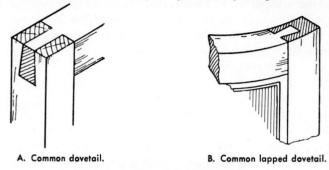

A. Common dovetail.   B. Common lapped dovetail.

Fig. 23. Two typical dovetail joints.

necting fixed shelves to partitions, because the dovetails prevent the partitions from bending. A dovetailed and housed joint, frequently called a "diminished" dovetail, is shown in Fig. 24D;

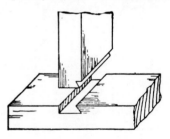

A. Barefaced dovetail housing.

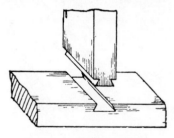

B. Common housed dovetail.

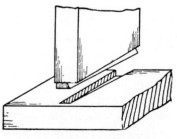

C. Shouldered housing dovetail.

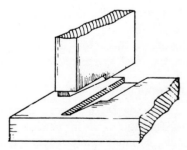

D. Dovetailed and housed.

Fig. 24. Housed dovetails of the single- and double-shouldered types.

it is principally used on comparatively small work, such as small fixed shelves and drawer rails.

Making a Diminished Dovetail— Square division lines across the ends into which the shelf is to be housed and dovetailed as far apart as the thickness of the shelf, and gauge the depth of the housing on the back edge. Gauge lines $3/8$ and $4\frac{1}{2}$ inches from the front edge between the division lines; the space between these gauge lines is the length of the actual dovetail, as shown in Fig. 25A.

Cut out the section indicated at **A** with a chisel, and undercut side **B** to form a dovetail; insert a tenon saw, and cut the sides across to the edge. Remove the core with a firmer chisel, and finish to depth with a router. Gauge lines on both ends of the shelf on the side and end for the depth of the dovetail, and square across the end the distances from the front edge as given; cut away the surplus wood with a tenon saw, as shown in Fig.

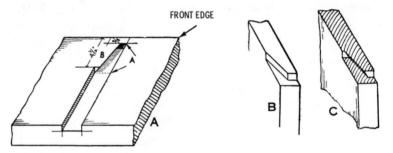

Fig. 25. The method of making a diminished dovetail joint.

25B, and finish the cut with a chisel, carefully testing until it fits hand tight. Fig. 25C shows the completed end. The average length of the actual dovetail of this type is slightly less than $\frac{1}{4}$ of the total length of dovetail and housing.

Mitered Joints

Mitering is an important part of cabinetwork in the framing of furniture and in panelling, where many difficult mouldings must be mitered into place. Fig. 26A illustrates a plain miter with a cross tongue (or spline) inserted at right angles to the miter. This joint is principally used for mitering end grain and is addi-

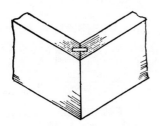

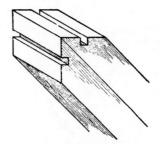

A. A tongued miter.   B. Method of tonguing a miter.

Fig. 26. Mitered joints.

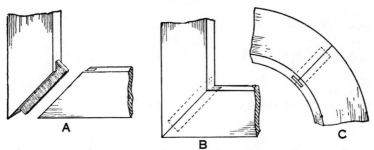

Fig. 27. The tongued miter joint in various stages and types of construction; A, before completion; B, completed joint; C, for connecting segments in curved work.

tionally strengthened by gluing a block to the internal angle, as shown in Fig. 29.

Tonguing a Miter—One practical way of tonguing this joint is to fasten two miters together in a vise so as to form a right angle, as in Fig. 26B, thus providing an edge from which to gauge the position of the tongues and plow the grooves. If the pieces are not over 6 inches in width, the grooves are cut with a dovetail saw and chiselled to depth.

Fig. 27A is a variation of plain mitering and, like the preceding joint, is most generally used for end-grain jointing. For this joint, the tongue should be approximately ⅓ the thickness of the material, and it may extend all the way through or only part way, as shown in Fig. 27B. This joint is especially useful in cabinetwork for connecting and mitering various types of large mouldings around the tops of pieces, for mitering material for

319

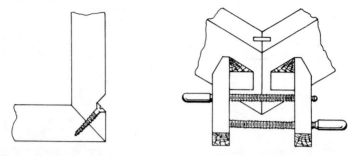

A. Cross-sectional view of the completed joint.

B. Method of clamping a miter joint.

Fig. 28. The screwed miter joint.

tops and panels, and for connecting sections in curved work, as shown in Fig. 27C.

Screwed Miter Joint—Fig. 28A illustrates a plain miter with a screw driven at right angles to the miter across the joint through a notch cut in the outside of the frame. This type of joint is used principally in light moulded frames.

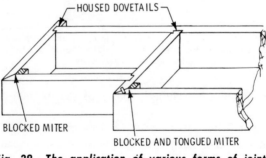

Fig. 29. The application of various forms of joints in cabinet construction.

A common method of clamping a tongued miter is to glue blocks to the piece and hand-screw the joint together, as shown in Fig. 28B. The blocks are glued on and allowed to dry before gluing up the joint; when the joint is dry, the blocks are knocked off, and their marks are erased.

Fig. 29 is a part plan for the base of a break-front cabinet; it shows the application of mitered and housed dovetail joints to furniture construction.

Framing Joints

The term "framed," or "framing," as used in cabinetwork, indicates work that is framed together, as in Fig. 29. It also refers to the "grounds" for securing panelling to walls.

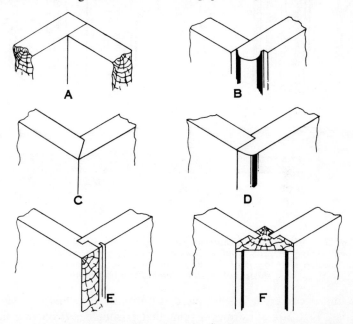

Fig. 30. Commonly used framing joints; A, butt, or square, joint; B, return bead and butt joint; C, rabbet and miter joint; D, rabbet and round joint; E, barefaced tongued joint; F, splayed corner joint.

Fig. 30 represents various joints that are used to connect angles for panelling. The joints shown in Fig. 30A and B are identical except for the return bead, which is worked on one of the pieces. These two joints are usually glued and nailed, but, when used as an external angle, they are secret-screwed before painting; that is, the screws are sunk below the surface, and plugs or pellets of wood are glued in the holes and beveled off. The joint represented in Fig. 30C may be used to connect framing at any angle; the rabbet prevents slipping while being nailed or screwed. Fig. 30D shows an ordinary rabbeted joint with the corner rounded off and the pieces glued together; because of its

321

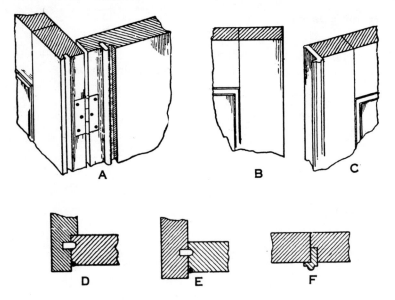

Fig. 31. Hinging and shutting joints commonly employed in cabinet construction; A, rabbeted dustproof joint; B, astragal shutting joint; C, beaded shutting joint; D, plain hinged joint; E, plain butted hinged dustproof joint; F, beaded shutting joint.

rounded corner, it is often used in furniture for children's nurseries. Fig. 30E illustrates a joint that is shouldered on one side only. A bead is worked on the tongue piece to hide the joint. It is used for both internal and external angles, with or without the bead. The joint in Fig. 30F is the splayed corner tongued joint, which is used for joining sides into a pilaster corner.

Hinging and Shutting Joints

The dustproof joint and its applications are considered a necessary part of cabinetmaking. The rabbeted dustproof joint, shown in Fig. 31A, is applied to a butt-hinged double door closet. A bead is glued into the rabbeted end behind the hinge, which is sunk flush, and a corresponding groove is cut in the door stile to fit over the bead when the door is closed. The beads are sometimes covered with felt or rubber, thereby making the joint absolutely dustproof.

322

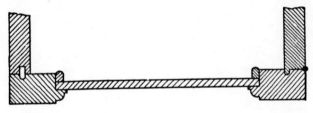

Fig. 32. A single-door dustproof joint. Both ends are dustproof when the door is closed.

Fig. 31B and C illustrate the shutting joint with a beaded strip glued to the stile; the joint is rabbeted to project over and thus conceal any shrinkage in the adjoining stile. The joint at the hinged ends (when closed) is shown in Fig. 31D. Fig. 31E represents a plain butted hinged dustproof joint, and Fig. 31F shows the beaded shutting joint when closed but with the astragal rabbeted into the stile.

SUMMARY

A great variety of joints used in cabinetwork are usually classified according to their general characteristics, such as glued, halved and bridle, mortise-and-tenon, dovetail, mitered, framing, and hinging and shutting. In cabinetwork, all joints should be glued. All glued joints must be clamped, or pressure applied by means of nails or screws.

The glue which seems to be the most popular in woodworking shops is the polyvinyls. This glue does not stain the wood, it is light in color, easy to use, and can set up at room temperature. Good glued joints must be in intimate contact while they are setting up. If this contact is not made, the joint will not have the strength required to hold up under normal service.

REVIEW QUESTIONS

1. Name a few of the tools used in cabinetmaking joints.
2. Why is it so important to clamp glued joints
3. What type of glue is best for general glued joints?
4. What is a coopered joint?
5. What is a dovetail joint?

CHAPTER 23

Wood Patternmaking

The term "patternmaking," as applied here, means the making of patterns to be used in forming molds for castings. Patterns are made of different materials, such as metal, plaster, and wood. Most but not all patterns are made of wood. Patterns for every known kind of machine, from the smallest to the largest, are made of wood. Therefore, this chapter is confined to the subject of wood patternmaking.

Wood patternmaking is regarded as the most highly skilled branch of the carpentry trade because of the greater amount of skill and technical knowledge required. It includes joiners' work and the mastery of woodworking tools and machinery, in addition to a knowledge of wood carving and turning, the ability to read complicated blueprints and visualize the shape and form of the pattern from the print, and a thorough understanding of foundry and core work.

Although a pattern is defined as a model, its outward appearance does not always closely resemble the casting itself, except in a simple casting where the pattern is often a complete model. If, however, the casting is to have interior passages and external openings, its appearance will be changed by the addition of projections called "core prints." These are placed so as to form bearing surfaces in the mold to support the sand cores used to form those passages or openings in the castings. These cores are usually formed in wooden molds called "core boxes," and the prints on the pattern are commonly distinguished by being painted a different color than the pattern itself. A pattern is given a certain amount of taper, or "draft," to insure its easy removal from the molding sand; the removing process is called "drawing." A pattern is said to draw well or not draw well according to the amount of trouble experienced in removing it from the molding sand.

PATTERNMAKER'S TOOLS

The first requisite of a patternmaker is a complete set of good tools. The following list will be found adequate for a wide range of work:

1. Jack plane.
2. Block plane.
3. End-wood plane (for planing long end-wood edges; it is 14 inches long, with the iron ground straight along the cutting edge, often referred to as a "jack plane").
4. Jointer plane.
5. Rabbet plane.
6. Router.
7. Circular plane (adjustable to both concave and convex surfaces).
8. Core-box plane.
9. Paring chisel (Fig. 1A).
10. Paring gouge (Fig. 1B).
11. Outside ground gouge (so called because it is ground on the outside, or convex, side; it is shorter and heavier than the paring gouge and is used for roughing work and with the mallet when necessary).
12. Carving tools (those most commonly used are known as straight, short-bend, and long-bend tools; a number of different sweeps of each style).
13. Bit brace (ratchet).
14. Hand drill.
15. Automatic drill.
16. Auger bits (in sizes from $\frac{3}{16}$ to 1 inch).
17. Gimlet bit (for drilling screw holes).
18. Center bit (for drilling thin stock).
19. Forstner bit (for drilling flat-bottom holes).
20. Expansion bit.
21. Bit stock drills (standard twist drills with square shank to fit hand braces, $\frac{1}{16}'' - 1\frac{1}{4}''$).

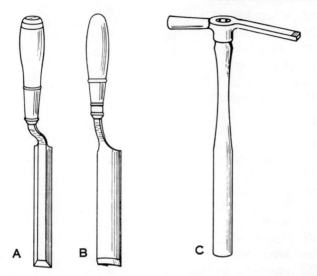

A   B   C

Fig. 1. The paring chisel (A), the paring gouge (B), and the pattern-maker's hammer (C) are three of the many tools used in patternmaking. The chisel varies in width from 1/8 to 2 inches, with square or beveled edges and straight or offset tangs; for general use, the 1/2-inch width is the most convenient, preferably with beveled edges and an offset tang, as illustrated. The paring gouge, like the chisel, is long and thin and varies in width from 1/8 to 2 inches, having three different curve sweeps known as "flat," "middle," and "regular." They are called inside-ground gouges, because they are ground on the inner, or concaved, side. The patternmaker's hammer is especially designed to meet the requirements of the trade; the long slender end is used for driving nails in fillets and for reaching corners that cannot be reached with the ordinary hammer.

22. Screwdriver bit.
23. Countersink bits.
24. Bit stop.
25. Back saw.
26. Keyhole saw.
27. Coping saw.
28. Hammer (patternmaker's) (Fig. 1C).
29. Claw hammer (medium size).
30. Oilstones (and slip stones of different shapes for sharpening the edge of cutting tools).
31. Spokeshave (large and small).
32. Hand screws.

33. Pinch-dogs (Fig. 2).
34. Bar clamps.
35. Distance marking gauge (monkey gauge).
36. Panel gauge.
37. Turning tools.
38. Hermaphrodite calipers.
39. Outside calipers.
40. Inside calipers.
41. Dividers (large and small).
42. Trammels (with inside and outside caliper points).
43. Bevel square.
44. Combination square.
45. Screwdrivers.
46. Scribers.
47. Shrink rules.
48. Shop machinery—power-driven machine tools of various types and numbers are standard equipment for the modern patternmaking shop; these woodworking machines make it possible to do work rapidly and accurately. The equipment includes:

    a. Circular saw.
    b. Band saw.
    c. Jointer.
    d. Surfacer.
    e. Trimmer.
    f. Jigsaw.
    g. Sanding machines.
    h. Disc sander.
    i. Roll sander.
    j. Wood milling machine.
    k. Tool grinder.
    l. Lathe.

TRADE TERMS

There are certain terms in common use in the patternmaking shop and in the foundry with which the prospective patternmaker should become familiar in order that he may have an intelligent

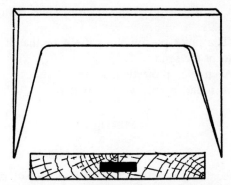

Fig. 2. A typical pinch-dog. The taper on the inside of the legs causes them to "pinch" when driven into two adjoining pieces.

understanding of his work. The most essential of these, as shown in Fig. 3, are:

Flask—A wood or metal frame made in two parts.
Cope—The top part of the flask.
Drag—The bottom part of the flask.
Bottom, or Mold, Board—A board which lies under the drag on which the pattern rests while the mold is being

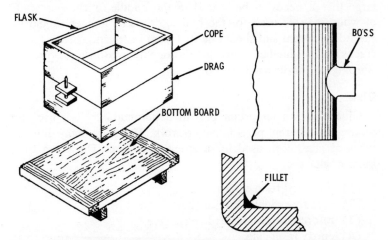

Fig. 3. The patternmaker's flask, with its components, and the boss and fillet, all of which are used in the patternmaker's trade.

329

rammed. It is also used on the top while rolling the mold over.

Draft—The taper put on the pattern so that it will "draw" easily out of the sand.

Boss—A circular projection, knob, or stud.

Fillet—A round corner used in patterns.

MATERIALS

Lumber

The most common varieties of wood used in patternmaking are white pine, mahogany, cherry, maple, and birch. White pine is considered by far the best wood for all but the smallest patterns. It is soft and therefore easy to work; porous enough to take glue well, thereby insuring strong glue joints; and, when properly seasoned, is not greatly affected by exposure to heat, cold, or dampness. Cherry wood is used when greater strength is required in the pattern. It is also the wood most generally used for small patterns. Mahogany is used for patterns of light, thin construction that require a great amount of strength and hand working. Maple and birch are too hard for economical hand tool work, but they are well suited for small turned patterns. When a pattern is so large that it needs to be braced in the middle, a cheaper grade of wood may be used for the bracing. Pattern lumber should be carefully selected and should be free from knots and shakes (small cracks). The lumber must be thoroughly seasoned to prevent warping of the finished pattern.

Glue

Glue plays an important part in patternmaking; it is used for uniting the different parts in patternwork. There are many different types of glue, each used for a specific purpose. The seven basic kinds of glue are:

(1) polyvinyl resin emulsion glue,
(2) urea formaldehyde resin glue,
(3) resorcinol-formaldehyde resin glue,
(4) epoxy resin glue,
(5) contact cement,

(6) casein glue,
(7) animal glue.

The surfaces of a joint should be made perfectly true before applying the glue. The surfaces that make contact should be dry, clean, and smooth prior to the spreading of the glue and the surface should be thoroughly and evenly coated. Although the application is thorough, it should not be heavy. A heavy coat is messy and wasteful.

Shellac or Pattern Varnish

Yellow shellac (commercially known as orange shellac) is used in pattern shops for varnishing finished patterns as a protection against atmospheric moisture and the wet molding sand, which would warp them, and to make the patterns draw easier from the molding sand.

Pattern Colors

It is a general shop practice to indicate core prints and core-box faces by some given color and to paint the patterns with some recognized code of colors to correspond to the metals in which they are to be cast. For example:

Pattern and core-box bodies for iron casting are painted black.
Patterns and core-box bodies for steel casting are painted blue.
Patterns and core-box bodies for brass casting are painted orange.
All core prints and core-box faces are painted red.

Colored shellac is used for this purpose; it is made by dissolving the powdered pigment in alcohol and mixing it with the orange shellac.

Dowel Pins

Some patterns must be put together in such a way that certain parts can be easily removed and replaced again as required; a split pattern is a typical example. Small dowel pins of wood or metal are used for this purpose to assure proper alignment.

Wooden dowel pins are used if the pattern is to be employed only a few times, but where there is much wear expected on a pattern and also on large patterns, metal dowels and dowel plates are invariably used. Two styles of metal dowel plates commonly used are shown in Fig. 4. They are secured with screws to the two parts of the pattern.

Fillets

Fillets, or concave connecting pieces, are used in the corners and at the intersection of surfaces of a pattern for the important reasons that they increase the strength of a casting by influencing

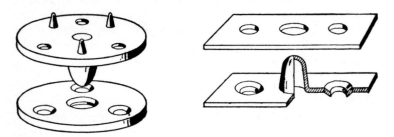

Fig. 4. Two types of metal dowel plates.

the crystallization of the metal and also improve the appearance of the casting. Fillets are of two types—"stuck" and "planted." A stuck fillet is one which is cut with a gouge out of the wood left on the pattern for that purpose, as shown in Fig. 5A. A planted fillet is one made separately and fitted in. Planted fillets are most commonly used and are made of wood, leather, and wax (beeswax).

Wood fillets are used for corners having large radii and on straight work. A round plane and a fillet board, Fig. 5B and C, are used for planing wood fillets. The stock is first reduced to a triangular section (Fig. 6A) on the circular saw and cut to suitable lengths; it is then put on the fillet board and planed with the round plane. A projecting screw head at the end of each groove acts as a stop. Wood fillets are always glued in place and are then securely nailed. To prevent the edges from curling, wet the face or concave side of the fillet with water before applying the glue. Thin strips are usually tacked over the shim, or feather, edges on large fillets, as in Fig. 6B, to hold them down until the glue is set, after which they are removed.

Leather fillets are strips of leather which are triangular in shape and are usually furnished in 4-foot lengths, as shown in Fig. 6C. They are extremely pliable and are easily attached to straight or curved work or to sharp or round corners, as shown in Fig. 6D. If moistened with water, their pliability is increased. They are fastened with glue or thick shellac and are rubbed in place with a waxing iron (Fig. 6E). If glued, the rubbing must be done rapidly and before the glue sets. The use of shellac per-

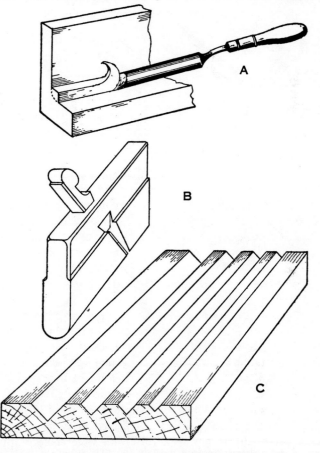

Fig. 5. Various fillet-shaping tools; A, cutting a fillet with a gouge; B, round plane; C, fillet board.

mits more time for rubbing. Coat the back of the fillet as well as the corner to which it is to be applied; let the shellac get sticky before rubbing.

Beeswax fillets can be used on small patterns that are used infrequently but not if the pattern is likely to be molded in sand which has become warm enough to melt the wax. The fillets are

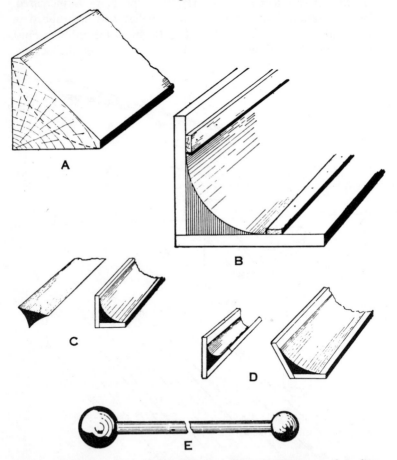

Fig. 6. Methods used to obtain different fillet shapes; A, stock is first reduced and then cut, as indicated by the dotted line, with a round plane; B, application of wood fillets; C, leather fillets and their appliciaton; D, leather fillets may readily be attached to sharp or rounded corners; E, typical waxing iron used in the application of leather and beeswax fillets.

applied after the first coat of varnish. The wax is prepared for use as fillets with a beeswax gun. The wax is forced out through an opening in the side of the gun in the form of a long string that

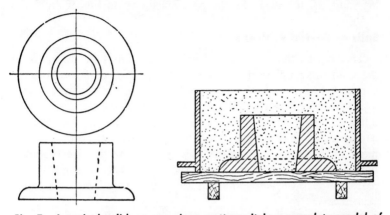

Fig. 7. A typical solid, or one-piece, pattern. It is a complete model of the required casting, including the control hole, which is large enough to let the molding sand stand in it as a green sand core to form the hole in the casting. The pattern is placed on the molding board with the larger side down and the drag part of the flask placed over it. The surface of the pattern is covered with facing or fine molding sand, which is rammed in and around it, and the drag is filled with coarser sand, which is rammed flush with the top. A bottom board is placed on top of the drag, which is then turned over; the first board is then removed. Parting sand is dusted over the exposed side of the drag; the cope side of the flask is put in place and is rammed flush with the top.

is ready for use; it is rubbed into the corners with the waxing iron.

TYPES OF PATTERNS

Generally speaking, patterns are divided into two classes—solid or one-piece patterns and split or parted patterns. In addition, there are some special types, such as skeleton patterns, part patterns, and patterns with loose pieces.

Solid or One-Piece Patterns

A solid or one-piece pattern is made without partings, joints, or any loose pieces; it is of one-piece construction when finished, and it may be a complete model of the required casting, or it may be partly cored, as shown in Figs. 7 and 8. Solid patterns require core prints for coring central holes when the hole is of such a size or shape that it cannot leave its own core in the molding sand, as, for example, the patterns shown in Fig. 9.

Split or Parted Patterns

This is one that, because of its shape, cannot be drawn from the sand mold unless it is made in at least two parts, such as an engine cylinder with its flanges, shown in Fig. 10. Such a pattern will lie half in the drag, or lower part of the flask, while the other half lifts off with the cope, or upper part of the flask. It is important to keep the parts directly over each other in the mold to insure a true casting; this is done by the use of dowel pins, as shown. Split patterns should, whenever possible, be parted and doweled together on the joint or parting of the mold.

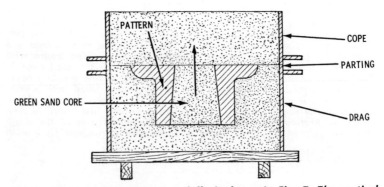

Fig. 8. The one-piece pattern and flask shown in Fig. 7. The vertical arrow indicates the direction in which the pattern is to be drawn from the mold.

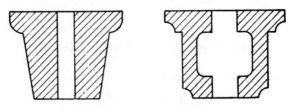

Fig. 9. Typical one-piece patterns.

Skeleton Patterns

Skeleton patterns are wooden frames that are used in framing a portion of the pattern in sand or clay. After the sand or clay is rammed inside the frame, it is worked to the required form by pieces of wood called "strickles."

Part Patterns

Part patterns are sections of a pattern that are so arranged as to form a complete mold by being moved to form each section of the mold. The movements are guided either by following a line or by the use of a central pin or pivot. These patterns are generally applied to circular work.

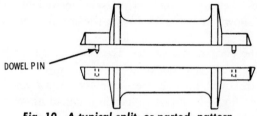

DOWEL PIN

Fig. 10. A typical split, or parted, pattern.

Patterns With Loose Pieces

Patterns with loose pieces are patterns with projecting parts that form undercuts; these undercuts prevent the loose pieces from being drawn with the main body of the pattern without destroying the mold. The undercutting projections are made as loose pieces and are fastened in place with loose dowels, or "skewers." By removing the skewers, the pattern is freed; the loose pieces which remain in the mold when the pattern is drawn are drawn into the cavity left by the pattern and are removed from the mold; this procedure is called "picking in."

CORES, CORE PRINTS, AND CORE BOXES

Dry sand core is a molded form that is made of sand mixed with a binder and baked until dry and firm. It is placed in the mold to form the interior passage, or opening, in the casting. "Green sand core" is the term used where the sand is allowed to stand in the mold and the metal is permitted to run around it to form the interior passage, or opening, in the casting.

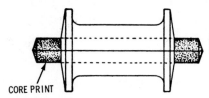

CORE PRINT

Fig. 11. A split, or parted, pattern and its core print are used for making a hollow casting.

Core Prints

When a casting is to be made hollow, its pattern must be made with a core print, such as the one shown in Fig. 11. This core print leaves a cavity or shelf in the sand mold into which is laid a core, as shown in Fig. 12A. When the metal is poured into the mold, it surrounds the core and leaves an opening of the size and shape of the core in the casting. The sand core is easily knocked out after the metal has cooled.

There are two kinds of core prints—vertical and horizontal. Core prints should be given a taper in order to more readily draw them from the sand, especially in the case of vertical core prints. The usual core-print taper is approximately ⅛ inch per inch. Horizontal cores must usually be supported at both ends, as shown in Fig. 12A. The length of the core prints at these points is decided by the weight of the core; a heavy core requires longer prints to give sufficient support to the cores and to guard against the crumbling of the sand-mold edges. For vertical cores, the lower core print should be longer than the upper one, since the lower core print practically supports the entire core; the upper core print serves only to keep the core from moving, as shown in Fig. 12B.

Core Boxes

A core is molded to the required shape and size in a core box, Fig. 12C. When both halves of a core are exactly alike, this type of half core box is used to save the patternmaker's time. The core is made in two halves, which are pasted together after they are baked. If it is necessary to make a complete circular core box, the two parts are usually held together by dowels, thus making it

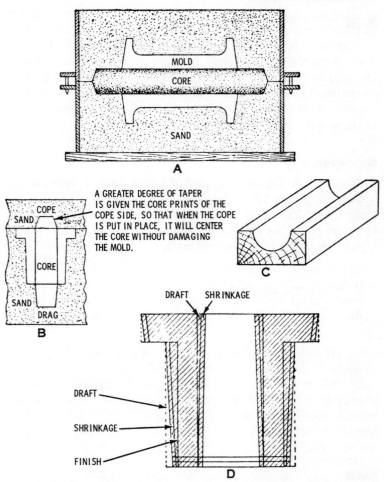

Fig. 12. Various methods used to obtain a hollow casting; A, horizon-tal-core method; B, vertical-core method; C, core box; D, draft, shrink-age, and finish allowances required on a pattern for a vertical casting.

easy to remove the core from the box. In a core box for a large pattern, some allowance for expansion should be made, since cores expand in baking. If the cores are not dried too quickly, the figure generally used for core-box allowance is $\frac{1}{16}$ inch per foot. To in-sure good surfaces on the core, make sure that the face of the core box is smooth.

Draft

A taper, or "draft," must be given to vertical parts of a pattern; if this were not done, it would be impossible to draw the pattern from the molding sand without damaging the mold. This draft is shown in Fig. 12D. The size and shape of a pattern determines the amount of draft to be allowed. However, it is a good practice to allow approximately ⅛ inch per foot of length.

Shrinkage

When the molten metal is poured into a mold, it contracts as it cools and leaves the casting smaller than its mold. Therefore, in order for the casting to be the required size, this shrinkage must be allowed for when making the pattern. The shrinkage allowances usually employed by patternmakers are:

Iron	⅛ inch per foot
Steel	³⁄₁₆ inch per foot
Brass	³⁄₁₆ inch per foot

Special rules, called "shrinkage rules," are available for the patternmaker's use. These rules contain the shrinkage of the casting metal, thereby eliminating the necessity of determining the proper allowance for the metal. In appearance, they look the same as an ordinary rule except that they are longer. Thus, a 2-foot shrink rule for ⅛ inch shrinkage per foot will be ¼ inch longer than the standard rule.

Finish

When any of the surfaces of a pattern are to be machined and finished off, an allowance for the finish must be made in addition to that for draft and shrinkage, as shown in Fig. 12D. The amount allowed for this varies from approximately ¹⁄₁₆ inch to ¼ inch, according to the location and nature of the casting, the methods of machining it, and the degree of finish required.

A good general rule to remember for the allowance on cast iron, small or medium sized work to be finished in a lathe, planer, or milling machine, is approximately ⅛ inch per foot. For larger pieces, the allowance will vary from ¼ to ¾ inch. An extra finish allowance is usually made on the cope side of patterns for large castings to permit machining to the sound metal beneath the slag, which always rises to the top of a mold. If a casting is to be finished on only one end, the pattern should be marked so as to make that fact known to the foundry man. Some foundries paint the finished surfaces green.

Blueprints

A blueprint is the plan of a casting from which the pattern-maker is required to construct a pattern. On standard blueprints, there is a title block in the lower right-hand corner. It gives the company's name and the draftsman's name. It tells whether the drawing is full size, half size, or quarter size and also describes the kind of material to be used for the casting.

The notes on the blueprint should be carefully studied. When a drawing is to be scaled for dimensions, a standard scale should always be used on the work. If the dimension is marked ¾ *inch,* it means that the drawing is out of scale. If holes are marked *drill,* this means the casting will be drilled. If the blueprint is marked *bored* or with an "*f*" (indicating finish), an allowance should be left on the pattern for this; the amount of the allowance is determined by the kind of material to be used for the casting. On small work where the blueprint is marked *spot face* or *disc grind,* only about 1/32 inch for the finish allowance is added.

JOINTS AND JOINERY

Joinery is defined as the art of a joiner. It is, in fact, the art of joint making and as such is an important part of patternmaking. It is work which requires the utmost skill because of the precision and accuracy with which the parts must be shaped and fitted together. Joinery involves planing straight edges on the parts to be joined together; facing or forming true or plane surfaces on the sides of flat stock; cutting grooves, mortises, and dovetails; planing, sawing, or trimming edges to angles; fitting parts to circular or irregular forms; and many other similar operations.

There are various forms of joints used in the construction of patterns and core boxes, and it is essential to know the different forms and when and where to use them to the best advantage. They are:

1. Straight or butt.
2. Checked.
3. Half-lapped.
4. Tongue and groove.
5. Splined.
6. Rabbet.
7. Dado.
8. Mortise and tenon.
9. Dovetail.

Butt Joint

The butt joint, Fig. 13A, is the simplest form of joint and is probably used most frequently of all, especially for light framework. It is not too strong even when properly glued and should be reinforced whenever possible by nails, screws, or dowels.

Checked Joint

This joint, Fig. 13B, is also known as the housed, butt, or rabbeted joint. It has an advantage over the butt joint in that it may be nailed or screwed from the edge. It is also adaptable to the formation of corner fillets.

Half-Lapped Joint

There are three varieties of this joint in general use—the corner lap, the cross lap, and the center lap. The center lap is sometimes dovetailed, Fig. 13C, D, E, and F. The half-lapped joint is considered the best all-around fastening joint for light frames. When properly glued and fastened with screws, it makes a strong joint that is commonly used in the joining of ribs and webs.

Tongue-and-Groove Joint

This form of joint, Fig. 13G, is often used for fortifying butt joints between ribs, etc. and for plate work in the making of boards for mounting patterns.

Splined Joint

The joint in Fig. 13H is the same as the feather joint or the plowed-and-tongued joint. It serves the same purpose as the tongue-and-groove joint, but it is preferred when joining two pieces of softwood because of the added strength given to the joint by the

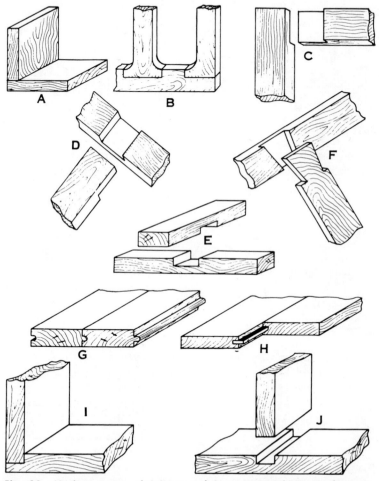

Fig. 13. *Various types of joints used in patternmaking; A, butt; B, checked or housed; C, half-lapped, corner; D, half-lapped, center; E, half-lapped, cross; F, half-lapped, center dovetailed; G, tongue-and-groove; H, splined; I, rabbet; J, dado.*

inserted spline, or feather. The spline is usually made of hardwood which is cut across the grain and fitted into grooves in the two joining pieces; this greatly reduces the liability of the joint to snap, as is often the case when the tongue is cut from softwood lengthwise with the grain.

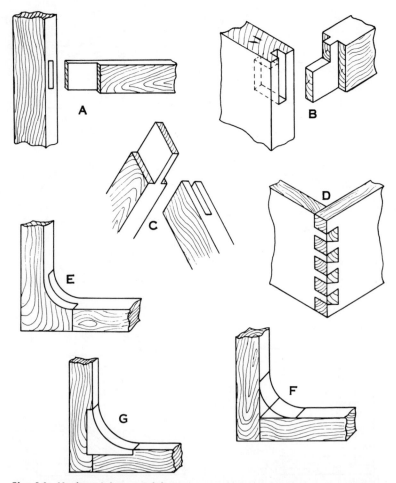

Fig. 14. Various joints used for square and curved patterns; A, mortise and tenon, through; B, mortise and tenon, blind; C, mortise and tenon, open; D, dovetail; E, half-lapped with corner fillets; F, butt with two-piece fillet; G, checked or housed fillet.

Rabbet Joint

The rabbet, or housed butt, joint, Fig. 13I, is easily made and is satisfactorily used on the sides and ends of core boxes.

Dado Joint

The dado joint, Fig. 13J, is quite satisfactory for fastening the ends of ribs in the sides of frames and boxes.

Mortise and Tenon Joint

This joint, though occasionally used with good results in some pattern framework, is scarcely used at all in small pattern work. This joint may be either through, blind, or open, as illustrated in Fig. 14A, B, and C.

Dovetail Joint

This joint, Fig. 14D, is exceptionally strong and is used in places which are difficult to fasten with screws or nails, such as the corners of light beds and open-sided boxes. It is also used to some extent on loose pieces for first-class permanent patterns and core boxes.

Squaring Framework

To square large frames, use a measuring rod placed diagonally from corner to corner; move the sides of the frame until the two diagonals are equal. On large box work, use the steel square, and check with the rod. Nail a batten to the face to hold it square until the glue sets and also until the corner fillets, if any, have been fastened in place.

Corner Fillets on Thin Framework

On thin frames, corner fillets are frequently made by leaving sufficient stock on the ends of the frame from which they are shaped after the half-lapped joint is made, as shown in Fig. 14E. If the fillets are of large radii, they are usually put in separately because of the cost involved. They are often made of two pieces, as in Fig. 14F, and sometimes they are made in one piece and are checked or "housed," as in Fig. 14G. They are fastened roughly in place by gluing and nailing and then worked into shape with a paring gouge.

Hardwood Corners and Edges

The corners and edges of pine patterns which are likely to be subjected to hard usage are reinforced with hardwood. Rabbets are cut in the pattern into which the hardwood pieces are fitted; in open boxes, the plowed-and-tongued, or splined, joint is frequently used to fasten the sides and ends to the corners. Ribs are often edged with hardwood. The inside corners of boxed work and the outside corners of core boxes are reinforced with corner blocks of hardwood which are glued and nailed in place.

Joints for Loose Pieces

On all first-class work, loose pieces that are to be left in the mold when the pattern is drawn are dovetailed in place. A dovetail is cut on the pattern, and a piece to fit it (forming the pin) is fastened to the back of the loose piece, as shown in Fig. 15A. When laying out the dovetail, provide an ample taper in the direction of the draw, as shown in Fig. 15B, and plane the pin approximately $\frac{1}{2}$ inch longer and slightly thicker than the depth of the dovetail from the face to the bottom. The additional length allows for cutting off the stock at the bottom if the dovetail should be cut too large, and the added thickness permits the dovetail piece to be planed flush with the pattern face. A back saw is used to cut the dovetail socket; cut as close to the line as possible and as far as the lines will allow. The stock is removed from the socket with a chisel or finished to depth with a router.

Fillets on Loose Pieces

If fillets are required on loose pieces that are to be "picked in," the loose piece is gained, or grooved, into the pattern, as shown in Fig. 15C, in order to give the fillet a thicker and stronger edge, thus eliminating the feather edge, which is so easily broken. This thick edge is sometimes made in the form of a dovetail, as in the lower section of Fig. 15C.

346

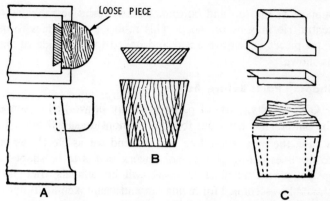

Fig. 15. Joint details for loose pieces; A, the loose piece; B, proper taper, in the direction of the draw, must be provided; C, the loose piece is grooved into the pattern to strengthen the fillet.

PATTERN DETAILS AND ASSEMBLY

Patternmaker's Box

The patternmaker's box, as shown in Fig. 16, is used as a foundation to make many patterns, especially if they have a regu-

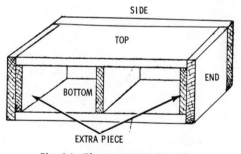

Fig. 16. The patternmaker's box.

lar outline and a rectangular form. The box should be made so as to have as few pieces of end wood as possible in contact with the draw sides. The top and bottom of the box are fitted between the sides and ends to prevent them from extending beyond or falling short of the sides through shrinkage or other causes. To add

347

support to the top and bottom, the ends may be rabbeted, and a central rib may be provided. This rabbet is sometimes formed by extra pieces which are glued and nailed to the inside of the box, as shown.

Shaping Parts Before Assembling

Occasionally, parts of patterns cannot be worked to shape after the pattern has been put together without great difficulty. In such a case, the parts should be laid out and cut as closely as possible to the lines before gluing. Framework that is to be shaped on the inside, where machine finishing will be difficult after assembly, should be assembled for laying out and then taken apart and sawed. Screws or dogs may be used to fasten the work while it is laid out. Pieces to be fitted in corners should be finished, even to sandpapering, before they are fastened in place.

Fitting to Cylindrical Forms

The usual procedure when fitting to cylindrical pieces is to chalk cylindrical parts and then rub and cut the piece to be fitted until

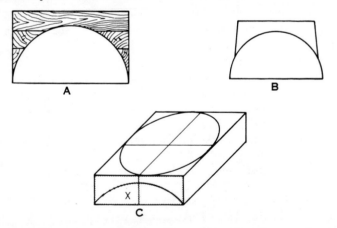

Fig. 17. The method of fitting a circular boss to a cylindrical body. The boss is laid out, as shown, and is sawed on the band saw to fit the cylinder; the waste piece X is used to hold the boss in position on the saw table while the boss is sawed to the required outline.

it forms a tight joint. This fitting is usually done on the band saw; some additional fitting afterwards may be necessary, but much time

348

should not be taken in rubbing and hand fitting to make a perfect joint, because the fillet covers the joint. When making bosses, use the thickest lumber available, and fit them as far as possible around the cylinder; fill in with one or more pieces after the boss is fastened, as shown in Fig. 17A. Fig. 17B and C illustrates an approved method of fitting a circular boss or branch to a cylindrical body.

Staved or Lagged Work

Staved work is one of three methods used in constructing patterns and core boxes of cylindrical shape. The other two methods are known as stepped work and segment work. A staved, or lagged, pattern is constructed by fastening barrow strips, called "staves" or "lags," to foundation pieces, called "heads." Fig. 18A shows the staves fastened to the heads of a pattern for one-half of a regular cylinder that is to be parted lengthwise through the center. Large cylindrical work that is to be finished either by hand or by turning is usually constructed by this method; it provides the maximum amount of strength and makes it possible to construct close to the finished outline of the pattern.

The staves are ripped in narrow pieces and beveled on each side from stock previously planed to an even thickness; they are then fastened to the heads by means of glue and nails. The number and thickness of staves depends largely on the size of the job (they are usually about 1 inch thick). However, the amount of gluing surface between the staves is important if a good joint is desired; it should not be less than $3/4$ inch.

In medium-size work, the undersides of the staves are often concaved to fit the circle of the head by passing them over the circular saw at an angle. The saw is projected above the table at a distance equal to height X in Fig. 18B, and one of the staves, with the narrow side up, is clamped on the saw table at an angle, as shown in Fig. 18C. This angle is determined by squaring off the stave so that the distance from the edge of the stave to the front of the saw is equal to the width of the narrow side of the stave, as at Y. If the radius required on the stave is greater than the radius of the saw, two or three cuts may be necessary on each stave to cut to the radius line. When a large radius is used, the heads are cut flat under the staves; the circumference is spaced in even parts, and the staves are cut to fit, as in Fig. 18D. A brace, or rapping and lifting bar, is run through each head at the parting

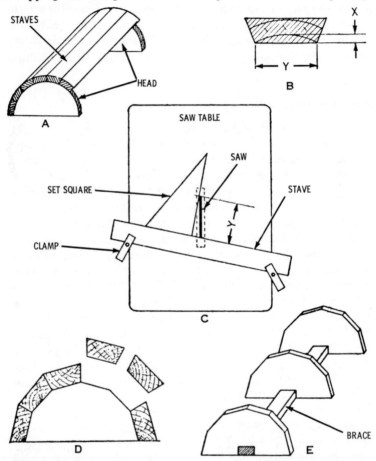

Fig. 18. The staved-work method of making cylindrical patterns; A, staves fastened to the heads of a pattern for half of a regular cylinder; B, cutting the staves; C, staves are clamped and sawed on a saw table; D, staves for a large radius cylinder; E, a brace is used to strengthen the pattern.

line, or center of the pattern, to strengthen the pattern by tying the heads together and to provide the means for rapping and lifting in the foundry, as shown in Fig. 18E.

The assembly of the staves and heads is done on a perfectly smooth, straight board having a straight edge. A center line is drawn parallel to the straight edge, and lines representing the length of the pattern are squared from this edge across the board. The heads are set to these lines with their center lines matching

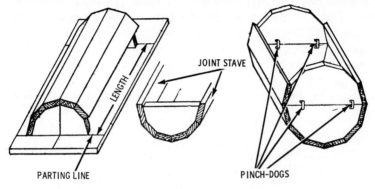

Fig. 19. Assembly of staved patterns.

the center line on the board, as in Fig. 19A; each stave is then glued and nailed in place, beginning at the center or parting line and working to the center or top of each half.

Some patternmakers begin construction by fastening the second stave from the bottom in place first and then continuing until the second stave from the bottom on the opposite side is in place. This is done because the joint staves are left slightly wider than the others, as shown in Fig. 19B, to allow for hand planing of the joint.

If the pattern is to be a complete cylinder, then the second half of each head should be doweled to the first and fastened together with pinch-dogs driven in both sides (inside and outside) of each head, as shown in Fig. 19C. Fasten the joint staves on each side of this half section first, and continue building from these. Before fastening the last stave in place, remove the dogs on the inside of the heads.

Staved Core Boxes

The principles involved in the construction of staved core boxes are the same as for staved patterns, only in reverse; the staves are finished on the inside and the outside and are also braced or supported. The ends of staved boxes are called "heads" and are usually connected by braces at the face and bottom, as shown in Fig. 20A. Round heads and concaved staves are used for long boxes, as shown in Fig. 20B. The end heads are depended on to keep the box together, because the outside bracing may or may not be added. The box is built on a board, the same as a cylinder. If the box is exceptionally long, several heads may be used and then removed after the box is put together. This construction produces a true box, which is finished when the staves are put on.

Stepped work gets its name from the fact that the stock, when fastened together, resembles steps. It is the method used for con-

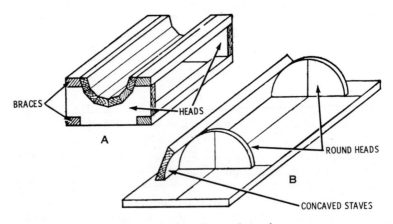

Fig. 20. Assembly of staved core boxes.

structing cylindrical forms, such as straight and curved pipes, up to 8 or 9 inches in diameter that are to be finished in the lathe, and for elbows and bends of all sizes that are to be finished by hand.

Fig. 21 illustrates how a parted cylinder pattern is constructed by the stepped method. A layout is made of one-half of the cylinder and the core print, as shown in Fig. 21A. The lumber used should be planed to thickness, since the number of steps depends on the thickness, and the two top pieces of the stock and the different steps should be cut to width according to the measurements taken

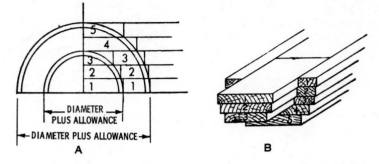

Fig. 21. The construction of a parted-cylinder pattern by the stepped method.

on the layout. The outside diameter is divided into equal parts on the vertical center line, and horizontal lines are then drawn through these points. Perpendicular lines, which are erected at the intersections of the circles and the horizontal lines, as shown, will give the width of the top pieces and the different steps.

The steps are assembled on the top pieces, as shown in Fig. 21B: the ends are marked so as to locate each piece. They are then glued and held in place with clamps or pinch-dogs. When gluing pieces such as these to a line, toenail each end to prevent the piece from shifting while applying the clamps or pinch-dogs. The pieces for the prints are fitted after the glue has dried. The joint step should be thicker than the others to provide for truing up the face on the jointer.

Segment Work

The primary purpose of segment work is to have the grain of the wood follow the outline of the pattern as far as is possible, because this procedure provides greater strength and makes it easier to finish the pattern, especially if the pattern is to be turned, since most of the end wood is eliminated. Segment work is used for the construction of curved ribs and similar pieces. When a complete circle is to be constructed, successive layers of wood, called "courses," are built up to the thickness of the required piece; these courses are divided into an equal number of divisions, called "segments," as shown in Fig. 22A. The number of courses for a given job depends largely on the thickness of the stock available and also

353

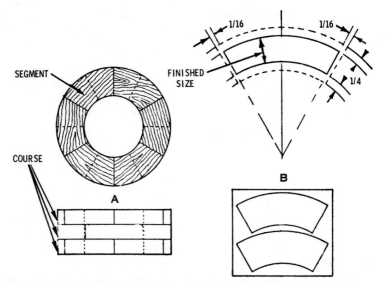

Fig. 22. The method of obtaining a cylindrical pattern by the segment method.

on how this stock corresponds with the total thickness required. At least three rows of courses are necessary to avoid warping. An easy way to determine the number of courses required is to make a layout of a small section of the job, adding the necessary amount of allowance for the finishing operation, and divide this to suit the lumber.

Fig. 22B shows the layout for a segment which is to be used as a pattern to mark the outline on the stock, as shown at **X**, having six segments to each course. When there are six segments to a course, the length of each one is equal to the radius of the circle. An allowance of $\frac{1}{16}$ inch is added to each end of the segment for fitting. The $\frac{1}{4}$-inch allowance is cut off after the segments have been laid up.

The band saw should be used when sawing segments; cut as close to the lines as possible. The ends of the segments are cut to the correct angle on the trimming machine, which is equipped to give the correct angle set for any number of segments in a course from 3 to 12. The different courses in segment work are glued

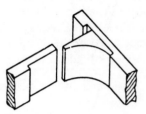

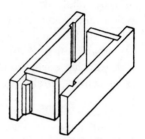

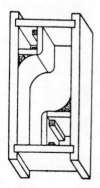

together with the segment joints of one course coming halfway between the joints of the course above and below. Thick glue is used and should be well rubbed into the segment ends to size them. Reinforce the glue by nailing whenever possible.

Core-Box Construction

The common method of making a core box for square or rectangular cores is to have the ends dadoed into the sides, with the box capable of being parted at diagonal corners to release the core, as shown in Fig. 23A. The loose corners are fastened with

A. Diagonal parting corners.   B. Filleted corners.

Fig. 23. Core-box construction.

screws, around the heads of which are drawn circles with black varnish to indicate that they are to be removed to part the box and release the core. The dadoes on the loose corners are usually made more shallow than those on the fastened corners in order to make the box part freely. Corner blocks, securely glued and nailed in

Fig. 24. The use of fillers to make the core shorter, narrower, or both, than the box.

place, are used to reinforce the fastened corners. If corner fillets are required, they are glued in place at the fastened corners; at the parting corners, they must be made a part of the side of the box, as shown in Fig. 23B.

Fillers—Blocks known as fillers are placed in rectangular core boxes to make the core shorter or narrower, or both, than the box and to change the outline of the box, as shown in Fig. 24. Fillers are fastened in the box or placed loosely in position as the occasion requires.

Round Core Boxes—The term "round core box" usually means a half-round box. To lay out a box of this type, plane the stock true on the face and on one edge; square the ends, and cut to length. There should be a margin on the face of the box along each side of the required diameter of not less than 1 inch. A distance equal to this margin is gauged from the working edge; from the pieces bradded to each end of the stock as centers, circles are

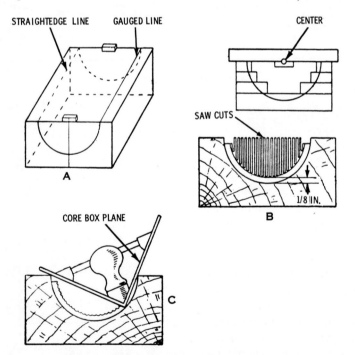

Fig. 25. The construction of round core boxes.

drawn on each end which coincide with the gauged line on the face, as shown in Fig. 25A. Connect the opposite sides of the circles by a line which has been either gauged or drawn with a straightedge. The circles are laid out on the ends of stepped and lagged boxes by the use of the device shown in Fig. 25B.

Core boxes made from solid timber are roughed out by making a series of cuts with the circular saw up to within $\frac{1}{8}$ inch of the radius line, as shown in Fig. 25C, and then removing the sawed stock with a gouge. On stepped or lagged boxes, the roughing-out procedure is done with an offset-tang gouge. A core-box plane is used to remove the remaining stock. To start the plane, a rabbet approximately $\frac{1}{16}$ inch deep is planed on both sides of the box by fastening a thin strip of wood to the face of the box along the layout lines as a guide for the plane. Remove the strip and plane from the right side to the center, as in Fig. 25D. Remove the stock, start from the right, and plane to the center again. Sand until smooth with No. $\frac{1}{2}$ sandpaper and a mandrel at least $\frac{1}{16}$ inch smaller than the size of the desired core.

SUMMARY

Wood patternmaking is a most highly skilled branch of the carpentry trade. It takes a great amount of skill and knowledge to develop this technical trade. It requires a trade in jointer work and woodworking tools and machinery, in addition to a knowledge of wood carving and turning. A person must master the ability to read blueprints and to visualize the shape and form of the pattern from a diagram or blueprint.

A complete set of tools are required in patternmaking, such as various sizes and types of planes, chisels, carving tools, hand and automatic drills, bits, saws, clamps, and squares. Electric machinery such as drills, sanders, saws, and lathes makes it possible to do work rapidly and accurately.

Various cabinetmaking joints, as discussed in Chapter 25, are required in patternmaking, such as butt joints, miter joints, and dado. The most common wood used in patternmaking are white pine, mahogany, cherry, maple, and birch.

Staved work is one of three methods used in constructing patterns and core boxes of cylindrical shape. The other two methods are known as stepped work and segment work. A staved pattern is constructed by fastening barrow strips, called staves, to foundation pieces, called heads. The staves are fastened to the heads by means of glue and nails.

REVIEW QUESTIONS

1. What are the most common woods used in patternmaking?
2. What is a patternmakers box?
3. What is a stave? What is a head?
4. What are the advantages of power-driven machine tools?
5. What is a pinch-dog?

CHAPTER 24

Kitchen Cabinet Construction

Proper planning of the kitchen layout constitutes one of the more important phases of modern living. A great amount of study has been made in order to provide proper storage space in the form of cabinets and equipment. These should be properly placed so as to reduce to a minimum the wasted motions and steps necessary in the preparation, cooking, and serving of food.

The size and number of cabinets required depends on the kitchen area available, the amount of cooking to be done, and the distance of the home from store centers. Thus, a home that is located within a short distance of stores will not need the storage space required in a home that is remotely located from shopping centers, nor would a small family need the same amount of storage space as that required by a large family.

When remodeling a kitchen, there are certain limitations to be observed. For example, the locations of the sink and range are usually predetermined, thus making it necessary to plan, construct, and install the various cabinets and equipment so as to keep these two centers in their original locations.

Once the layout has been decided on, the next step consists of choosing the number and type of cabinets. The cabinets are usually designed so that they can be made as individual units and installed as such, or several cabinets can be made and installed as a combination. When a combination of cabinets is to be arranged to form a large unit, it is usually better to make a large top of sufficient length to cover the two or more units placed side by side than to make a separate top for each unit.

The number and the location of electrical outlets and fixtures are other important parts of kitchen planning. Wall outlets should

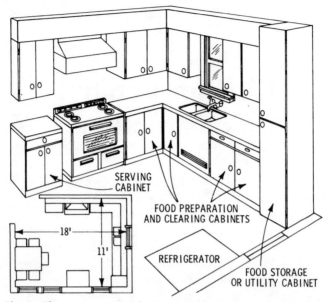

Fig. 1. The arrangement of storage cabinets in a typical kitchen.

be placed above the floor cabinets for easy connection to electrical appliances which are used in the preparation of food, such as toasters, food mixers, coffeemakers, casseroles, roasters, broilers, etc. It is considered good practice to have several double outlets available to permit the maximum utilization of labor-saving electrical appliances. Modern electrical- and gas-equipped kitchen ranges are commonly furnished with permanently installed fluorescent lamps, which prevent shadow problems when working at the kitchen range. To provide kitchen ventilation and to remove cooking odors, an exhaust fan installed in an outside wall is another useful appliance.

CONSTRUCTION CONSIDERATIONS

Although it may readily be admitted that not every carpenter has sufficient skill to design and make really artistic cabinets suitable for kitchen use, there are a great many simple cabinets which can be easily made. In general, it is best to start out with a cabinet of simple construction, and when greater proficiency has been attained in the handling of tools and the working of wood, the more intricate and difficult projects may be undertaken.

KITCHEN WALL CABINETS

These, as the name implies, are designed to be hung on the wall above the floor cabinets and are suitable companion pieces to the floor units. As noted in Fig. 2, they are rather simple in construction and may easily be made with the simple carpenters' tools which are available in most homes. The design illustrated permits construction in widths ranging from 16 to 36 inches, depending on the available and desired space.

For the most convenient usage, it has been found that cabinets of this type should be proportioned so that the top of the cabinet will be approximately 7 feet above the floor surface. Cabinets ranging from 16 to 24 inches in width usually require only one door, while cabinets of 28 to 36 inches in width require two doors. It should be noted, however, that regardless of size, all cabinets are similar in construction; the only difference is in the width and the height. When the width and height of the cabinet have been decided on, the material for each cabinet should be ordered. As the different pieces are cut, they should be marked with a key letter to facilitate assembly. If the cabinet is to be constructed of solid wood rather than plywood, it may be necessary to glue up two or more narrow widths of stock to produce panels having the required widths.

The actual working drawings shown in the accompanying illustrations are based on a cabinet which is 28 inches wide, although the construction procedure will be similar for cabinets of any width.

Construction Details

When making a cabinet such as illustrated, it should be noted that actual construction is started with the side members. These pieces, as shown in the illustration, have a rabbet cut on the upper end to take the top piece, a rabbet cut along the back edge to take the back panel, a dado cut near the lower end for the bottom member, and two grooves cut at the upper and lower back edges to take the back rails.

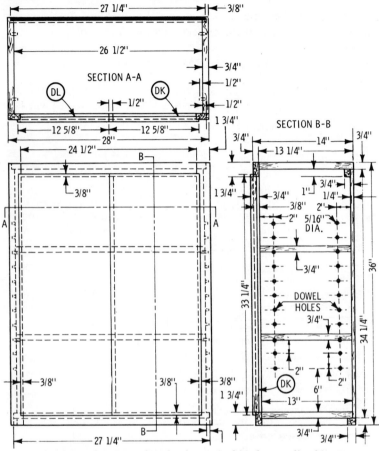

Fig. 2. *Construction details of a typical kitchen wall cabinet.*

After the necessary work on the two side members, the carcass for the cabinet can be assembled. When the carcass is completely assembled, the corners should be checked with a try square to make certain that they are square. A temporary diagonal brace may be fastened to the front edge to keep the unit square while the back panel is being applied.

The front frame is constructed of the two stiles and the bottom and top rails. The rails are joined to the stiles by means of mortise and tenon joints. Glue is spread over the joints, and the frame is set in clamps. The corners of the assembled frame must be checked to make certain that they are square. The shelves are cut to size in the regular manner, but they require the cutting of small circular grooves in each end to properly engage the supporting dowels. The door, or doors, should not cause any difficulty when assembling them to the unit if they are properly cut to fit the cabinet.

To complete any cabinet, hardware of some type is required. The types of cabinet hardware available are almost unlimited, as will be noted by catalog descriptions. Various classes of hinges, door pulls, catches, and latches are generally found at any well-equipped hardware store. If catches are to be placed on double doors, then the strikes should be fastened to the underside of the shelf, to the top rail of the cabinet above the door, to the lower rail below the door, or to the bottom shelf. The strike and catch are usually provided with elongated holes in order to permit their adjustment after they have been installed. If a series of cabinets is being planned, the necessary hardware required may conveniently be

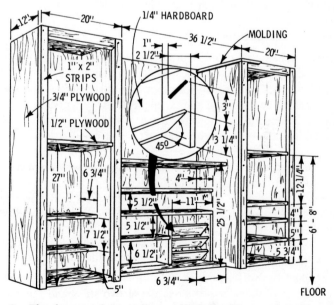

Fig. 3. The layout of a wall-type kitchen-cabinet unit with center canned-food racks. Metal angles on the wall are used to support the cabinet.

purchased at one time, thus assuring the perfect matching of each unit.

Cabinet Installation

Because of the considerable weight of a fully loaded cabinet, such as the combination unit shown in Fig. 3, a great degree of care must be taken to insure its proper installation. In all homes of frame construction, as well as in many other types, the inside walls are constructed on studs, which are placed on 16- or 24-inch centers. These studs will, in most cases, provide the only supporting means for the cabinet. The studs may be located by various means, such as by a fine drill or an icepick; be careful not to damage the wall during the stud-locating process. A plumb line may be dropped from the ceiling at the center of the studs to obtain the center line of the studs throughout their entire length. Once the location of the studs has been ascertained, it is a simple matter to locate the holes in the top and bottom rails of the cabinet for final installation by means of suitable wood screws.

Use of a False Cabinet Wall

Most kitchen wall cabinets are designed so that the hanging wall units and full-length cabinets have a maximum height of 7 feet above the floor. The reason for observing this standard is to facilitate usage, since any shelf placed higher than this distance cannot normally be reached without the aid of a stepladder. With cabinets installed at standard heights, however, there will usually be an open space near the ceiling where dust will tend to accumulate. A convenient method of eliminating this dust-collecting space consists of constructing a false wall extending from the top of the cabinet to the ceiling. This method, when used, will add to the decor of the room and is well adaptable to one or several cabinets as required.

To enclose the space above the cabinets, a framework, such as shown in Fig. 4 must be installed. Wallboard or plywood can then be attached to this framework. The stock used for the framework need not be of heavier cross section than $1\frac{1}{4}'' \times 2\frac{1}{2}''$ or even lighter, depending on the bracing and height of the upright members. As noted in Fig. 4, cleats are fastened to the cabinet top and the ceiling to provide nailing strips for the uprights. Since the cleats are notched to receive the uprights, the panels may be fastened flush with the framework on all sides. Before nailing the cleats in place, however, the thickness of the panel must be determined, since the setback of the cleats which are attached to the cabinet top is controlled by this measurement. After having properly located the upper cleat, the next step is to locate the ceiling beams. If the ceiling beams run at right angles to the cleat, no difficulty will be encountered when fastening the cleat to the ceiling. If, on the other hand, the beams are found to run parallel to the cleat, the nailing strips will have to be applied at right angles to the beams and extended out far enough to provide a nailing surface for the cleats.

After the framework has been nailed in place, all that is necessary is to cut the panels to size. These are usually fastened by means of 4-penny nails. Scrollwork, when desired, is an optional feature. When applying the molding to the false cabinet wall, care should be taken to match the existing molding as closely as pos-

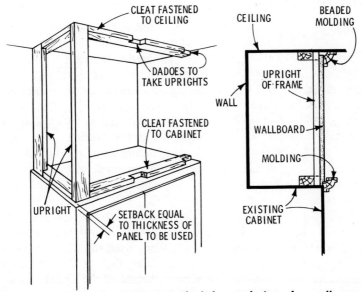

Fig. 4. A simple construction method for enclosing the wall space above an existing wall cabinet.

sible. These panels can usually be purchased at any well-equipped lumber yard.

The next step is the sanding and finishing operations. The sanding removes whatever scratches are left in the wood surfaces; the false cabinets are then painted to match the original cabinets. Also, if the paneled surface is to be papered, this work must be done before the molding is fastened.

BASE CABINET CONSTRUCTION

A typical kitchen cabinet of the base type is shown in Fig. 5. The cabinet is 24 inches deep from front to back and is 36 inches high. The cabinet, in most cases, is constructed of ¾ inch plywood. Although particle board and flake board is sometimes used. The rails and stiles are usually cut from ¾ inch plywood, although solid stock can be used. The sides of the base suit are dadoed to receive the shelf and plywood bottom. A 3½ × 3½ inch toe-space is provided at the bottom of the unit and 2 × 4 lookouts are placed under the plywood bottom. The sides of the base units are some-

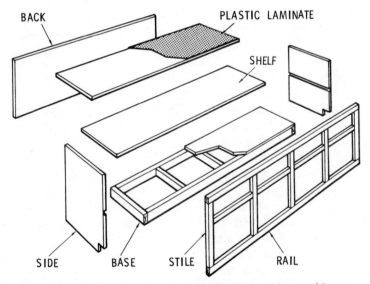

Fig. 5. Construction details of a typical kitchen unit base cabinet.

KITCHEN CABINET CONSTRUCTION

times rabbeted to receive a ¼ inch hardboard back or a 1 × 6 hanging strip is used in place of backing. All joints should be glued, using brads as required to anchor the work and prevent slipping.

Doors and Drawers

There are two basic types of drawers and doors that can be used, flush and lipped. The flush door fits between the rail and stile and is very difficult to fit. The lipped door has a ⅜ inch rabbet around the door, allowing for a margin of error. If a lipped door or drawer is used, the size of the front of the drawer or door is obtained by adding ½ inch to the size of the opening.

Plastic Laminate Tops

A high pressure plastic laminate top is an excellent counter top for a kitchen cabinet. The surface is functional as well as being decorative. Plastic laminate is not considered a structural material and should be backed by plywood or particle board. To cut the plastic laminate, use snips and a crosscut saw with fine teeth, a hacksaw, coping saw, or router.

For satisfactory results in attaching the plastic laminate to the cabinet top, use an adhesive which has been recommended by the manufacturers of the laminate. Several manufacturers of laminated plastic suggest the use of a contact-bond adhesive which contains good adhesive properties. The adhesive is first spread on the back of the laminate with a brush. As soon as the laminate has been fully covered, spread the adhesive on the cabinet top. The adhesive is then allowed to dry until a piece of wrapping paper will not stick to the adhesive. The edge of the counter top is first covered with the plastic laminate. Once the blind edge is covered it should be routered or filed flush with the cabinet top.

After the edge of the counter top is finished, the top of the cabinet surface is covered with a piece of clean wrapping paper. The paper is used as a slip sheet to assist in correctly placing the plastic. Put the plastic laminate on the cabinet top and make sure it is exactly in position with a small margin to spare on all sides. This procedure, if properly used, will be a great asset in the application process, because once the adhesive-covered surfaces touch one another no shifting of the plastic is possible. After the plastic laminate is positioned, the slip sheet is carefully removed. A rubber mallet or hand roller is then used to firmly bond the plastic laminate to the cabinet top. Once the cabinet topping is placed, a router is used to trim it to the proper size.

SUMMARY

The general number of cabinets required for the kitchen area depends on the size of the kitchen and the amount of cooking to be done. When remodeling a kitchen, there are certain limitations to observe; for example, the location of the sink and cooking range. The number and location of electrical outlets and fixtures are other important parts in kitchen planning.

Because of the considerable weight of a fully loaded cabinet, a great degree of care must be taken to insure proper fit at all joints. Most kitchen wall cabinets are designed so that the hanging wall units and full-length cabinets have a maximum height of 7 feet above the floor.

In most cases plastic laminated material is used for cabinet tops. This type of material is generally $\frac{1}{16}$ of an inch thick, resulting in a smooth surface. New counter tops should be constructed of $\frac{3}{4}$-inch exterior-grade plywood, which should be firmly fastened to the cabinet.

To complete any cabinet, hardware of some type is required. Various classes of hinges, door pulls, catches, and latches are generally found at any well-equipped hardware store.

REVIEW QUESTIONS

1. What determines the number of cabinets and the general placement?
2. What type material is installed as cabinet counter tops?
3. What type and grade of lumber is generally used in cabinet making?
4. What is generally done to the wall space above the wall-mounted cabinets?

Index

Chamfer plane, 180
Chisels, 155
  butt, 159
  corner, 158
  firmer, 156
  framing, 157
  gauge, 157
  mill, 159
  mortise, 157
  paring, 156
  pocket, 159
  slick, 157
  tang and socket, 159
Clamps, 127
Classifications of wood, 9-11
Combination square, 86, 93
Compass
  and divider, 107
  or keyhole saw, 135
Construction with steel square,
  rafters
    cripper, 246
    hip, 243
    jack, 245
    valley, 244
Coping saw, 136
Core-box construction, 355
Corner chisel, 158
Countersinks, 201
Crosscut saw, 139
Cypress wood, 20

D

Dado joint, 279, 345
Decay of wood, 23
Defects in wood, 14
Dimension of screws, 52
Doors and drawers, cabinet, 367
Double irons, plane, 184
Douglas fir wood, 22
Dovetail joints, 290, 314, 345
Dowel
  joints, 271, 304
  pins, 331
Drawknife, 165
Dressing
  handsaw, 153
  lumber, 13
Driving nails, 45
Drying lumber, 13

E

Eastern red cedar wood, 20

F

Fastening tools, 207-218
  hammer, 207
  screwdriver, 211
  wrenches, 214
Files and rasps, 141
Filing, handsaw, 149
Fillester plane, 179
Finding rafter length without table,
  246
Firmer chisel, 156
Fitting cylindrical forms, 348
Fore plane, 175
Framing
  chisel, 157
  joint, 231
  square, 88

G

Glued joints, 300
Gouge chisel, 157
Grinding
  tools, 225
  wheel, 219
Grooming plane, 179
Guiding and testing tools, 83-101
  level, 98
  miter box, 97
  plumb bob, 99
  square, 83
    combination, 86, 93
    framing, 88
    miter, 86
    sliding-T, 94
    steel, 88
    try, 84
  straightedge, 83
Gum red wood, 20

H

Hacksaw, 136
Half-lapped joint, 342
Halved and bridle joint, 307
Hammer, 207
Hand
  axe, 167
  drills, 202
Handsaw sharpening, 145
  dressing, 153
  filing, 149
  jointing, 146
  setting, 148
  shaping, 146

The Audel® Mail Order Bookstore

Here's an opportunity to order the valuable books you may have missed before and to build your own personal, comprehensive library of Audel books. You can choose from an extensive selection of technical guides and reference books. They will provide access to the same sources the experts use, put all the answers at your fingertips, and give you the know-how to complete even the most complicated building or repairing job, in the same professional way.

Each volume:
- **Fully illustrated**
- **Packed with up-to-date facts and figures**
- **Completely indexed for easy reference**

APPLIANCES

REFRIGERATION: HOME AND COMMERCIAL
Covers the whole realm of refrigeration equipment from fractional-horsepower water coolers, through domestic refrigerators, to multi-ton commercial installations. **Cat. No. 23286 Price: $8.95**

AIR CONDITIONING: HOME AND COMMERCIAL
A concise collection of basic information, tables, and charts for those interested in understanding, troubleshooting, and repairing home air conditioners and commercial installations. **Cat. No. 23288 Price: $7.50**

HOME APPLIANCE SERVICING, 3rd Edition
A practical book for electric & gas servicemen, mechanics & dealers. Covers the principles, servicing, and repairing of home appliances. **Cat. No. 23214 Price: $8.50**

REFRIGERATION AND AIR CONDITIONING LIBRARY—2 Vols.
Cat. No. 23305 Price: $15.95

OIL BURNERS, 3rd Edition
Provides complete information on all types of oil burners and associated equipment. Discusses burners—blowers—ignition transformers—electrodes—nozzles—fuel pumps—filters—controls. Installation and maintenance are stressed. **Cat. No. 23277 Price: $6.95**

Use the order coupon on the back page of this book.

AUTOMOTIVE

AUTO BODY REPAIR FOR THE DO-IT-YOURSELFER

Shows how to use touch-up paint; repair chips, scratches, and dents; remove and prevent rust; care for glass, doors, locks, lids, and vinyl tops; and clean and repair upholstery. Softcover. **Cat. No. 23238   Price: $5.95**

AUTOMOTIVE LIBRARY—2 Vols.
Cat. No. 23198   Price: $17.50

AUTOMOBILE REPAIR GUIDE, 4th Edition

A practical reference for auto mechanics, servicemen, trainees, and owners Explains theory, construction, and servicing of modern domestic motorcars. **Cat. No. 23291   Price: $11.95**

AUTO ENGINE TUNE-UP, 2nd Edition

This popular guide shows you exactly how to tune your car engine for extra power, gas economy and fewer costly repairs. **Cat. No. 23181   Price: $6.75**

CAN-DO TUNE-UP™ SERIES

Each book in this series comes with an audio tape cassette. Together they provide an organized set of instructions that will show you and talk you through the maintenance and tune-up procedures designed for your particular car. All books are softcover.

AMERICAN MOTORS CORPORATION CARS

(The 1964 thru 1974 cars covered include: Matador, Rambler, Gremlin, and AMC Jeep (Willys).) **Cat. No. 23843   Price: $7.95**
Cat. No. 23851 Without Cassette   **Price: $4.95**

CHRYSLER CORPORATION CARS

(The 1964 thru 1974 cars covered include: Chrysler, Dodge, and Plymouth) **Cat. No. 23825   Price: $7.95**
Cat. No. 23846 Without Cassette   **Price: $4.95**

FORD MOTOR COMPANY CARS

(The 1964 thru 1974 cars covered include: Ford, Lincoln, and Mercury.) **Cat. No. 23827   Price: $7.95**
Cat. No. 23848 Without Cassette   **Price: $4.95**

GENERAL MOTORS CORPORATION CARS

(The 1964 thru 1974 cars covered include: Buick, Cadillac, Chevrolet, Oldsmobile, and Pontiac.) **Cat. No. 23824   Price: $7.95**
Cat. No. 23845 Without Cassette   **Price: $4.95**

PINTO AND VEGA CARS,

1971 thru 1974. **Cat. No. 23831   Price: $7.95**
Cat. No. 23849 Without Cassette   **Price: $4.95**

TOYOTA AND DATSUN CARS,

1964 thru 1974. **Cat. No. 23835   Price: $7.95**
Cat. No. 23850 Without Cassette   **Price: $4.95**

VOLKSWAGEN CARS

(The 1964 thru 1974 cars covered include: Beetle, Super Beetle, and Karmann Ghia.) **Cat. No. 23826   Price: $7.95**
Cat. No. 23847 Without Cassette   **Price: $4.95**

Use the order coupon on the back page of this book.

DIESEL ENGINE MANUAL, 3rd Edition

A practical guide covering the theory, operation, and maintenance of modern diesel engines. Explains diesel principles—valves—timing—fuel pumps—pistons and rings—cylinders—lubrication—cooling system—fuel oil and more. **Cat. No. 23199 Price: $8.50**

GAS ENGINE MANUAL, 2nd Edition

A completely practical book covering the construction, operation, and repair of all types of modern gas engines. **Cat. No. 23245 Price: $7.95**

OUTBOARD MOTORS & BOATING, 3rd Edition

Provides the information you need to maintain, troubleshoot, repair, and adjust all types of outboard motors. Explains the basic principles of outboard motors and the functions of the various engine parts. Softcover. **Cat. No. 23279 Price: $6.95**

BUILDING AND MAINTENANCE

ANSWERS ON BLUEPRINT READING, 3rd Edition

Covers all types of blueprint reading for mechanics and builders. This book reveals the secret language of blueprints, step by step in easy stages. **Cat. No. 23283 Price: $6.95**

BUILDING A VACATION HOME

From selecting a building site to driving in the last nail, this book explains the entire process, with fully illustrated step-by-step details. Includes a complete set of drawings for a two-story vacation and/or retirement home. Softcover. **Cat. No. 23222 Price: $7.95**

BUILDING MAINTENANCE, 2nd Edition

Covers all the practical aspects of building maintenance. Painting and decorating; plumbing and pipe fitting; carpentry; heating maintenance; custodial practices and more. (A book for building owners, managers, and maintenance personnel.) **Cat. No. 23278 Price: $7.50**

GARDENING & LANDSCAPING

A comprehensive guide for homeowners and for industrial, municipal, and estate groundskeepers. Gives information on proper care of annual and perennial flowers; various house plants; greenhouse design and construction; insect and rodent controls; and more. **Cat. No. 23229 Price: $7.95**

CARPENTERS & BUILDERS LIBRARY, 4th Edition (4 Vols.)

A practical, illustrated trade assistant on modern construction for carpenters, builders, and all woodworkers. Explains in practical, concise language and illustrations all the principles, advances, and shortcuts based on modern practice. How to calculate various jobs. **Cat. No. 23244 Price: $24.50**

> Vol. 1—Tools, steel square, saw filing, joinery cabinets. **Cat. No. 23240 Price: $6.50**
>
> Vol. 2—Mathematics, plans, specifications, estimates. **Cat. No. 23241 Price: $6.50**
>
> Vol. 3—House and roof framing, laying out, foundations. **Cat. No. 23242 Price: $6.50**
>
> Vol. 4—Doors, windows. stairs, millwork, painting. **Cat. No. 23243 Price: $6.50**

Use the order coupon on the back page of this book.

CARPENTRY AND BUILDING

Answers to the problems encountered In today's building trades. The actual questions asked of an architect by carpenters and builders are answered in this book. **Cat. No. 23142   Price: $7.50**

HEATING, VENTILATING, AND AIR CONDITIONING LIBRARY (3 Vols.)

This three-volume set covers all types of furnaces, ductwork, air conditioners, heat pumps, radiant heaters, and water heaters, including swimming-pool heating systems. **Cat. No. 23227   Price: $25.50**

Volume 1

Partial Contents: Heating Fundamentals . . . Insulation Principles . . . Heating Fuels . . . Electric Heating System . . . Furnace Fundamentals . . . Gas-Fired Furnaces . . . Oil-Fired Furnaces . . . Coal-Fired Furnaces . . . Electric Furnaces. **Cat. No. 23248   Price: $8.95**

Volume 2

Partial Contents: Oil Burners . . . Gas Burners . . . Thermostats and Humidistats . . . Gas and Oil Controls . . . Pipes, Pipe Fitting, and Piping Details . . . Valves and Valve Installations. **Cat. No. 23249   Price: $8.95**

Volume 3

Partial Contents: Radiant Heating . . . Radiators, Convectors, and Unit Heaters . . . Stoves, Fireplaces, and Chimneys . . . Water Heaters and Other Appliances . . . Central Air Conditioning Systems . . . Humidifiers and Dehumidifiers. **Cat. No. 23250   Price: $8.95**

HOME MAINTENANCE AND REPAIR: Walls, Ceilings, and Floors

Easy-to-follow instructions for sprucing up and repairing the walls, ceiling, and floors of your home. Covers nail pops, plaster repair, painting, paneling, ceiling and bathroom tile, and sound control. Softcover. **Cat. No. 23281   Price: $5.95**

HOME PLUMBING HANDBOOK

A complete guide to home plumbing repair and installation. Softcover. **Cat. No. 23239   Price: $8.95**

HOME WORKSHOP & TOOL HANDY BOOK

Tells how to set up your own home workshop (basement, garage, or spare room) and explains the various hand and power tools (when, where, and how to use them). **Cat. No. 23208   Price: $6.50**

MASONS AND BUILDERS LIBRARY—2 Vols.

A practical, illustrated trade assistant on modern construction for bricklayers, stone-masons, cement workers, plasterers, and tile setters. Explains all the principles, advances, and shortcuts based on modern practice—including how to figure and calculate various jobs. **Cat. No. 23185   Price: $13.95**

Vol. 1—Concrete, Block, Tile, Terrazzo. **Cat. No. 23182   Price: $7.50**

Vol. 2—Bricklaying, Plastering, Rock Masonry, Clay Tile. **Cat. No. 23183   Price: $7.50**

Use the order coupon on the back page of this book.

PLUMBERS AND PIPE FITTERS LIBRARY—3 Vols.

A practical, illustrated trade assistant and reference for master plumbers, journeymen and apprentice pipe fitters, gas fitters and helpers, builders, contractors, and engineers. Explains in simple language, illustrations, diagrams, charts, graphs, and pictures, the principles of modern plumbing and pipe-fitting practices. **Cat. No. 23255 Price $19.95**

> Vol. 1—Materials, tools, roughing-in. **Cat. No. 23256 Price: $6.95**
>
> Vol. 2—Welding, heating, air-conditioning. **Cat. No. 23257 Price: $6.95**
>
> Vol. 3—Water supply, drainage, calculations. **Cat. No. 23258 Price: $6.95**

PLUMBERS HANDBOOK

A pocket manual providing reference material for plumbers and/or pipe fitters. General information sections contain data on cast-iron fittings, copper drainage fittings, plastic pipe, and repair of fixtures. Softcover. **Cat. No. 23246 Price $5.50**

QUESTIONS AND ANSWERS FOR PLUMBERS EXAMINATIONS,

2nd Edition

Answers plumbers' questions about types of fixtures to use, size of pipe to install, design of systems, size and location of septic tank systems, and procedures used in installing material. Softcover. **Cat. No. 23285 Price: $5.50**

TREE CARE MANUAL

The conscientious gardener's guide to healthy, beautiful trees. Covers planting, grafting, fertilizing, pruning, and spraying. Tells how to cope with insects, plant diseases, and environmental damage. Softcover. **Cat. No. 23280 Price: $8.95**

UPHOLSTERING

Upholstering is explained for the average householder and apprentice upholsterer. From repairing and regluing of the bare frame, to the final sewing or tacking, for antiques and most modern pieces, this book covers it all. **Cat. No. 23189 Price: $7.95**

WOOD FURNITURE: Finishing, Refinishing, Repairing

Presents the fundamentals of furniture repair for both veneer and solid wood. Gives complete instructions on refinishing procedures, which includes stripping the old finish, sanding, selecting the finish, and using wood fillers. **Cat. No. 23216 Price: $7.50**

ELECTRICITY/ELECTRONICS

ELECTRICAL LIBRARY

If you are a student of electricity or a practicing electrician, here is a very important and helpful library you should consider owning. You can learn the basics of electricity, study electric motors and wiring diagrams, learn how to interpret the NEC, and prepare for the electrician's examination by using these books. **Cat. No. 23316 Price $42.50**

Electric Motors, 3rd Edition. **Cat. No. 23264 Price: $7.95**

Guide to the 1978 National Electrical Code. **Cat. No. 23308 Price: $9.95**

House Wiring, 3rd Edition. **Cat. No. 23224 Price: $6.50**

Practical Electricity, 3rd Edition. **Cat. No. 23218 Price: $7.50**

Questions and Answers for Electricians Examinations, 6th Edition. **Cat. No. 23307 Price: $6.95**

Wiring Diagrams for Light and Power, 3rd Edition. **Cat. No. 23232 Price: $6.95**

ELECTRICAL COURSE FOR APPRENTICES AND JOURNEYMEN

A study course for apprentice or journeymen electricians. Covers electrical theory and its applications. **Cat. No. 23209 Price: $7.95**

Use the order coupon on the back page of this book.

RADIOMANS GUIDE, 4th Edition

Contains the latest information on radio and electronics from the basics through transistors. **Cat. No. 23259 Price: $7.50**

TELEVISION SERVICE MANUAL, 4th Edition

Provides the practical information necessary for accurate diagnosis and repair of both black-and-white and color television receivers. **Cat. No. 23247 Price: $7.95**

ENGINEERS/MECHANICS/ MACHINISTS

MACHINISTS LIBRARY, 2nd Edition

Covers modern machine-shop practice. Tells how to set up and operate lathes, screw and milling machines, shapers, drill presses, and all other machine tools. A complete reference library. **Cat. No. 23174 Price: $19.50**

Vol. 1—Basic Machine Shop. **Cat. No. 23175 Price: $6.75**

Vol. 2—Machine Shop. **Cat. No. 23176 Price: $6.75**

Vol. 3—Toolmakers Handy Book. **Cat. No. 23177 Price: $6.75**

MECHANICAL TRADES POCKET MANUAL

Provides practical reference material for mechanical tradesmen. This handbook covers methods, tools, equipment, procedures, and much more. Softcover. **Cat. No. 23215 Price: $4.50**

MILLWRIGHTS AND MECHANICS GUIDE, 2nd Edition

Practical information on plant installation, operation, and maintenance for millwrights, mechanics, maintenance men, erectors, riggers, foremen, inspectors, and superintendents. **Cat. No. 23201 Price: $11.95**

POWER PLANT ENGINEERS GUIDE, 2nd Edition

The complete steam or diesel power-plant engineers' library. **Cat. No. 23220 Price: $12.95**

QUESTIONS & ANSWERS FOR ENGINEERS AND FIREMANS EXAMINATIONS, 2nd Edition

An aid for stationary, marine, diesel & hoisting engineer's examinations for all grades of licenses. A new concise review explaining in detail the principles, facts, and figures of practical engineering. **Cat. No. 23217 Price: $7.95**

WELDERS GUIDE, 2nd Edition

This new edition is a practical and concise manual on the theory, practical operation, and maintenance of all welding machines. Fully covers both electric and oxy-gas welding. **Cat. No. 23202 Price: $11.95**

SHEET METAL WORKERS HANDY BOOK, 2nd Edition

Presents the fundamentals of sheet metal work layout in clear and simple language. **Cat. No. 23235 Price: $6.95**

Use the order coupon on the back page of this book.

FLUID POWER

PNEUMATICS AND HYDRAULICS, 3rd Edition

Fully discusses installation, operation, and maintenance of both HYDRAULIC AND PNEUMATIC (air) devices. Cat. No. 23237 Price: $7.95

PUMPS, 3rd Edition

A detailed book on all types of pumps from the old-fashioned kitchen variety to the most modern types. Covers construction, application, installation, and troubleshooting. Cat. No. 23292 Price: $7.95

HOBBY

COMPLETE COURSE IN STAINED GLASS

Written by an outstanding artist in the field of stained glass, this book is dedicated to all who love the beauty of the art. Ten complete lessons describe the required materials, how to obtain them, and explicit directions for making several stained glass projects. 80 pages; 8½ x 11; softbound. Cat. No. 23287 Price: $4.95

BUILD YOUR OWN AUDEL DO-IT-YOURSELF LIBRARY AT HOME!

Use the handy order coupon today to gain the valuable information you need in all the areas that once required a repairman. Save money and have fun while you learn to service your own air conditioner, automobile, and plumbing. Do your own professional carpentry, masonry, and wood furniture refinishing and repair. Build your own security systems. Find out how to repair your TV or Hi-Fi. Learn landscaping, upholstery, electronics and much, much more.

HERE'S HOW TO ORDER

Select the Audel book(s) you want, fill in the order card below, detach and mail today. Send no money now. You'll have 15 days to examine the books in the comfort of your own home. If not completely satisfied, simply return your order and owe nothing.

If you decide to keep the books, we will bill you for the total amount, plus a small charge for shipping and handling.

1. Enter the correct catalog number(s) of the book(s) you want in the space(s) provided.

2. Print your name, address, city, state and zip code, clearly.

3. Detach the order card below and mail today. No postage is required.

Detach postage-free order card on perforated line

FREE TRIAL ORDER CARD

☐ Please rush the following book(s) for my free trial. I understand if I'm not completely satisfied, I may return my order within 15 days and owe nothing. Otherwise, you will bill me for the total amount plus a small postage & handling charge.

Write book catalog numbers at right.

(Numbers are listed with titles)

NAME_____

ADDRESS_____

CITY_____ STATE_____ ZIP_____

☐ Save postage & handling costs. Full payment enclosed (Plus sales tax, if any.)

Cash must accompany orders under $5.00.
Money-Back guarantee still applies.

333

DETACH POSTAGE-PAID REPLY CARD BELOW AND MAIL TODAY!

Just select your books, enter the code numbers on the order card, fill out your name and address, and mail. There's no need to send money.

15-Day Free Trial On All Books . . .